164
Advances in Polymer Science

Springer-Verlag Berlin Heidelberg GmbH

Filler-Reinforced Elastomers Scanning Force Microscopy

With contributions by
B. Cappella · M. Geuss · M. Klüppel
M. Munz · E. Schulz · H. Sturm

Springer

The series presents critical reviews of the present and future trends in polymer and biopolymer science including chemistry, physical chemistry, physics and material science. It is addressed to all scientists at universities and in industry who wish to keep abreast of advances in the topics covered.

As a rule, contributions are specially commissioned. The editors and publishers will, however, always be pleased to receive suggestions and supplementary information. Papers are accepted for "Advances in Polymer Science" in English.

In references Advances in Polymer Science is abbreviated Adv Polym Sci and is cited as a journal.

The electronic content of APS may be found at http://www.SpringerLink.com

ISSN 0065-3195
ISBN 978-3-642-05604-8 ISBN 978-3-540-36438-2 (eBook)
DOI 10.1007/978-3-540-36438-2

Library of Congress Catalog Card Number 61642

http://www.springer.de

Originally published by Springer-Verlag Berlin Heidelberg New York in 2003
Softcover reprint of the hardcover 1st edition 2003

Typesetting: Stürtz AG, 97080 Würzburg
Cover: Design & Production, Heidelberg
Printed on acid-free paper 02/3020/kk - 5 4 3 2 1 0

Editorial Board

Advances in Polymer Science Available Electronically

For all customers with a standing order for Advances in Polymer Science we offer the electronic form via SpringerLink free of charge. Please contact your librarian who can receive a password for free access to the full articles. By registration at:

http://www.SpringerLink.com

If you do not have a standing order you can nevertheless browse through the table of contents of the volumes and the abstracts of each article by choosing Advances in Polymer Science within the Chemistry Online Library.

You will find information about the

- Editorial Board
- Aims and Scope
- Instructions for Authors
- Sample Contribution

at www.springeronline.com using the search function.

Contents

Adv Polym Sci (2003) 164:1–86
DOI 10.1007/978-3-540-36438-2

The Role of Disorder in Filler Reinforcement of Elastomers on Various Length Scales

Manfred Klüppel

Deutsches Institut für Kautschuktechnologie e. V., Eupener Strasse 33,
30519 Hannover, Germany
E-mail: Manfred.Klueppel@DIKautschuk.de

Abstract The chapter considers the disordered nature of filler networks on different length scales and relates it to the specific reinforcing properties of active fillers in elastomer composites. On nanoscopic length scales, the surface structure and primary aggregate morphology of carbon blacks, the most widely used filler in technical rubber goods, are analyzed by static gas adsorption and transmission electron microscopy (TEM) techniques, respectively. They are found to be closely related to two distinct disordered growth mechanisms during carbon black processing, surface growth and aggregate growth.

The role of disorder becomes also apparent on mesoscopic length scales of elastomer composites, where a filler network in formed due to attractive filler-filler interactions. An analysis of the d.c.-conductivity and dielectric properties of conductive carbon black-rubber composites indicates that no universal percolation structure is realized, but a superimposed kinetic aggregation mechanism of the particles takes place. The assumed kinetic cluster-cluster aggregation (CCA) of filler particles in elastomers is confirmed by the predicted scaling behavior of the small strain elastic modulus.

Based on the analysis of in-rubber morphology of filler particles and clusters on nanoscopic and mesoscopic length scales, a constitutive micro-mechanical model of stress softening and hysteresis of filler reinforced rubbers up to large strain is developed. It refers to a non-affine tube model of rubber elasticity, including hydrodynamic amplification of the rubber matrix by a fraction of rigid filler clusters with filler-filler bonds in the unbroken, virgin state. The filler-induced hysteresis is described by an anisotropic free energy density, considering the cyclic breakdown and re-aggregation of the residual fraction of more fragile filler clusters with already broken filler-filler bonds. Experimental investigations of the quasi-static stress-strain behavior of silica and carbon black filled rubbers up to large strain agree well with adaptations found by the developed model.

Keywords Elastomer composites · Disordered structures · Filler networking · Reinforcement · Micro-mechanics · Constitutive material laws

1 Introduction

Since the development of fractal geometry by Mandelbrot [1, 2] it has been learned that the apparently random structure of many colloidal aggregates formed in disorderly growth processes is subject to strong statistical constraints [3–8]. It is now well known that the fractal structure of colloidal aggregates results from the random movement, e.g., diffusion, of the aggregating particles that may be represented in computer simulations of cluster growth processes via random walks of single particles [3, 9] or whole clusters [10, 11]. Thereby, different universality classes of fractal aggregates can be distinguished by characteristic fractal exponents, dependent on the conditions of cluster growth [12].

In these studies we are primary interested in two classes of disorderly grown colloidal aggregates:

1. Clusters that are build in a diffusion limited cluster-cluster aggregation (CCA) process, where particles and clusters diffuse across each other and stick upon contact, irreversibly.
2. Clusters resulting from a ballistic cluster-cluster aggregation process, where particles and clusters move on linear trajectories.

A physical realization of the CCA-process is found during fast gelation of colloids in solution, where physical (and not chemical) bonds between particles and clusters are formed. From light scattering data of corresponding gold-, silica-, and polystyrene clusters as well as from computer simulations of CCA-clusters the fractal dimension takes the universal values $d_f \approx 1.8$ in all these cases [3–8, 12, 13]. The size distribution of the clusters is typically found to be broadly peaked around a maximum cluster size. This is in contrast to percolation clusters, where the size distribution shows a characteristic scaling behavior implying that the number of clusters increases successively with decreasing cluster size [6, 7].

A physical realization of ballistic cluster-cluster aggregation is found during carbon black processing, where surface growth due to random deposition of carbon nuclei also takes place [14–18]. Corresponding computer simulations of primary aggregate formation under ballistic conditions yields for the mass fractal dimension $d_f \approx 1.95$ [11, 12]. The surface growth of carbon black appears to be governed by a random deposition mechanism that falls into a universality class with a surface fractal dimension $d_s \approx 2.6$ [3, 6, 7, 19]. We will consider the simultaneous cluster and surface growth of carbon black more closely in Sect. 3, where the energetic surface structure is also analyzed [20].

Recently it has been argued that CCA-clusters are also built in elastomer composites. Accordingly, their particular structure can be used for the modeling of rubber reinforcement by active fillers like carbon blacks or silica [21–23]. This approach is not evident, because long range diffusion of filler particles or clusters, as assumed in the CCA-process, is strongly suppressed in high viscosity media. Instead, the polymeric structure of highly entangled rubbers give rise to fluctuations of colloidal particles around their mean position with a fluctuation length of the order of the entanglement length of the rubber [22–24]. For that reason the assumption of CCA-clusters in filler reinforced rubbers appears reasonable for sufficient large filler concentrations, only, where the mean trajectory length of aggregating particles or clusters becomes smaller than the fluctuation length. This condition is fulfilled for filler concentrations Φ above the gel point Φ^* of the filler network if the gelation concept of Ball and Brown is applied [25]. Then, the mean trajectory length in gelling systems becomes small and the cluster size is governed by the available empty space only. However, due to the restricted mobility of filler particles in rubber the gel point is shifted to filler concentrations that are orders of magnitude larger than the critical concentration Φ^* in solution [22, 23]. This is the main difference between the two systems that appears to be important for rubber reinforcement.

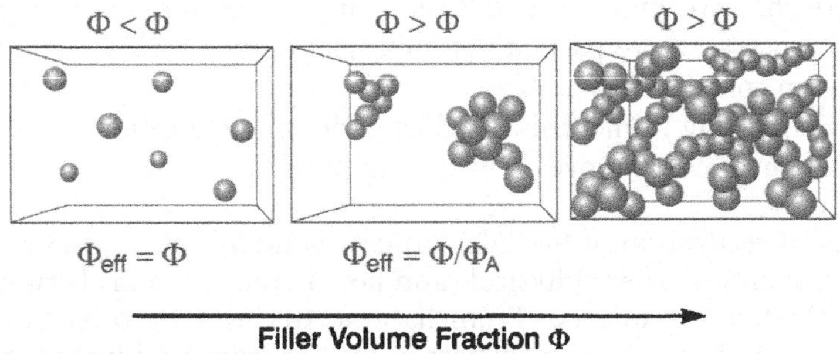

Fig. 1 Schematic view of filler morphology in three concentration regimes. For $\Phi<\Phi^*$ reinforcement is due to hydrodynamic amplification by particles ($\Phi<\Phi^+$) or clusters ($\Phi>\Phi^+$) with $\Phi_{eff}=\Phi$ or $\Phi_{eff}=\Phi/\Phi_A$, respectively. For $\Phi>\Phi^*$ reinforcement is due to the deformation of a flexible filler network

Different concentration limits of the filler arise from the CCA concept [22]. With increasing filler concentration first an aggregation limit Φ^+ is reached. For $\Phi>\Phi^+$, the distance of neighboring filler particles becomes sufficiently small for the onset of flocculation and clusters with solid fraction Φ_A are formed. Dependent on the concentration of filler particles, this flocculation process leads to spatially separated clusters or, for $\Phi>\Phi^*$, a through going filler network that can be considered as a space-filling configuration of fractal CCA-clusters. The different cases for spherical filler particles are shown schematically in Fig. 1.

So far, the formation and structure of filler networks in elastomers and the mechanical response, e.g., the pronounced dynamic amplitude dependence or stress softening, of reinforced rubbers, is not fully understood, though this question is of high technical interest. A deeper understanding of filler networking and reinforcement could provide a useful tool for the design, preparation, and testing of high performance elastomers, as applied in tires, seals, bearings, and other dynamically loaded elastomer components. Different attempts have been considered in the past that were primary focusing on the reinforcing mechanism of carbon black, the most widely used filler in rubber industry [26, 27]. The strongly non-linear dynamic-mechanical response of carbon black filled rubbers, reflected primarily by the amplitude dependence of the viscoelastic complex modulus, was brought into clear focus by the extensive work of Payne [28–35]. Therefore, this effect is often referred to as the Payne effect.

As shown in Fig. 2a, for a specific frequency and temperature, the storage modulus G' decreases from a small strain plateau value G'_0 to an apparently high amplitude plateau value G'_∞ with increasing strain amplitude. The loss modulus G'' shows a more or less pronounced peak. It can be evaluated from the tangent of the measured loss angle, $\tan\delta=G''/G''$, as depicted in Fig. 2b. Obviously, the loss tangent shows a low plateau value at small strain amplitude, almost independent of filler concentration, and passes through a broad maximum with increasing strain.

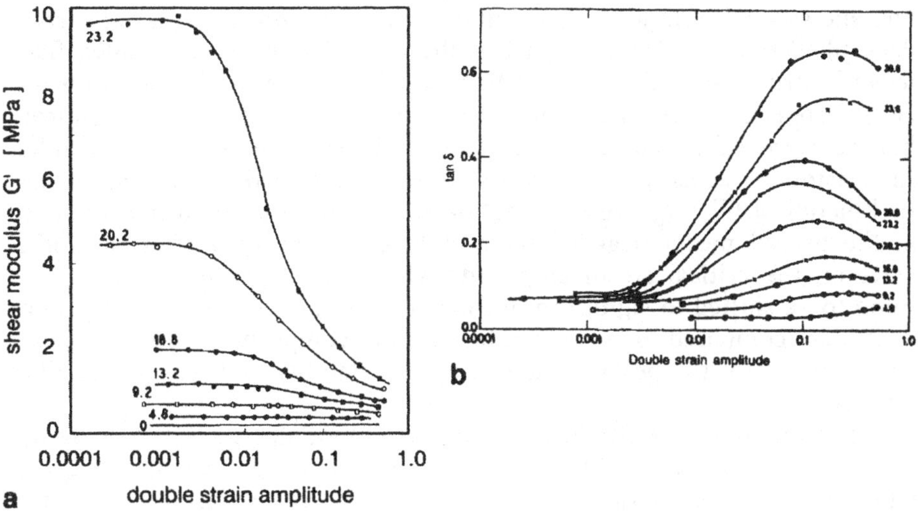

Fig. 2 a Amplitude dependence of the storage modulus of Butyl/N330-samples. b Amplitude dependence of the loss tangent of the same Butyl/N330-samples at various carbon black concentrations [28]

The Payne effect of carbon black reinforced rubbers has also been investigated intensively by a number of different researchers [36–39]. In most cases, standard diene rubbers widely used in the tire industry, like SBR, NR, and BR, have been applied, but also carbon black filled bromobutyl rubbers [40–42] or functional rubbers containing tin end-modified polymers [43] were used. The Payne effect was described in the framework of various experimental procedures, including pre-conditioning-, recovery- and dynamic stress-softening studies [44]. The typically almost reversible, non-linear response found for carbon black composites has also been observed for silica filled rubbers [44–46].

The temperature dependence of the Payne effect has been studied by Payne and other authors [28, 32, 47]. With increasing temperature an Arrhenius-like drop of the moduli is found if the deformation amplitude is kept constant. Beside this effect, the impact of filler surface characteristics in the non-linear dynamic properties of filler reinforced rubbers has been discussed in a review of Wang [47], where basic theoretical interpretations and modeling is presented. The Payne effect has also been investigated in composites containing polymeric model fillers, like microgels of different particle size and surface chemistry, which could provide some more insight into the fundamental mechanisms of rubber reinforcement by colloidal fillers [48, 49].

The pronounced amplitude dependence of the complex modulus, referred to as the Payne effect, has also been observed in low viscosity media, e.g., composites of carbon black with decane and liquid paraffin [50], carbon black suspensions in ethylene vinylacetate copolymers [51], and clay/water suspensions [52, 53]. It was found that the storage modulus decreases with dynamic strain amplitude in a qualitative manner similar to that of carbon black filled rubbers. This emphasizes the role of a physically bonded filler network structure in the Payne effect, which governs the small strain dynamic properties even in absence of rubber. Further, these results indicate that the Payne effect is primarily determined by structure effects of the filler. The elastomer seems to act merely as a dispersing medium that influences the kinetics of filler aggregation, but does not have a pronounced influence on the overall mechanical behavior of three-dimensional filler networks. However, the critical strain amplitude where the Payne effect appears is found to be shifted to significantly smaller values, if the low viscosity composites are compared to corresponding rubber composites. This indicates a strong impact of the polymer-matrix on the stability and strength of filler networks.

The strong non-linearity of the viscoelastic modulus with increasing dynamic strain amplitude has been related to a cyclic breakdown and re-aggregation of filler-filler bonds [21–23, 48, 54–57]. Thereby, different geometrical arrangements of particles in a particular filler network structure, resulting, e.g., from percolation as in the L-N-B-model of Lin and Lee [56] or kinetic cluster-cluster aggregation [21–23, 48], have been considered. Nevertheless, a full micro-mechanical description of energy storage and -dissipation in dynamically excited reinforced rubbers is still outstanding. A review of the different attempts is given by Heinrich and Klüppel [57].

Beside the Payne effect, relevant for dynamical loading of filler reinforced rubbers, the pronounced stress softening, characteristic for quasi-static deformations up to large strain, is of major interest for technical applications. It is often referred to as the Mullins effect due to the extensive studies of Mullins and coworkers [58–60] on the stress softening phenomena. Dependent on the history of straining, e.g., the extent of previous stretching, the rubber material undergoes an almost permanent change that alters the elastic properties and increases hysteresis, drastically. Most of the softening occurs in the first deformation and after a few deformation cycles the rubber approaches a steady state with a constant stress-strain behavior. The softening is usually only present at deformations smaller than the previous maximum. An example of (discontinuous) stress softening is shown in Fig. 3, where the maximum strain is increased, successively, from one uniaxial stretching cycle to the next.

So far the micro-mechanical origin of the Mullins effect is not totally understood [26, 36, 61]. Beside the action of the entropy elastic polymer network that is quite well understood on a molecular-statistical basis [24, 62], the impact of filler particles on stress-strain properties is of high importance. On the one hand the addition of hard filler particles leads to a stiffening of the rubber matrix that can be described by a hydrodynamic strain amplification factor [22, 63–65]. On the other, the constraints introduced into the system by filler-polymer bonds result in a decreased network entropy. Accordingly, the free energy that equals the negative entropy times the temperature increases linear with the effective number of network junctions [64–67]. A further effect is obtained from the formation of filler clusters or a

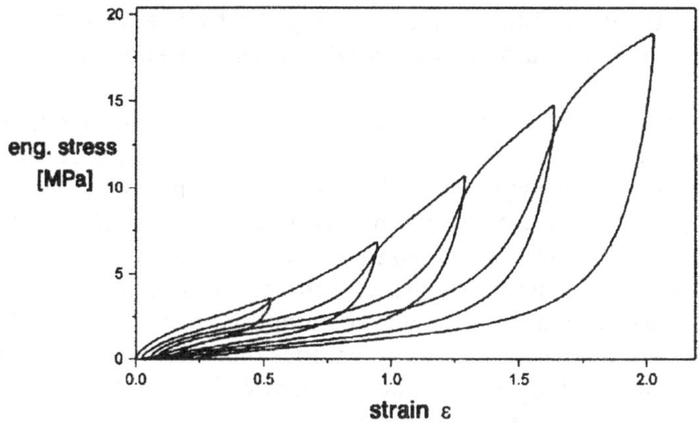

Fig. 3 Example of stress softening with successively increasing maximum strain for an E-SBR-sample filled with 80 phr N 339

filler network due to strong attractive filler-filler bonds [21, 22, 26, 36, 61, 64–67].

Stress softening is supposed to be affected by different influences and mechanisms that have been discussed by a variety of authors. In particular, it has been attributed to a breakdown or slippage [68–71] and disentanglements [72] of bonds between filler and rubber, a strain-induced crystallization-decrystallization [73, 74], or a rearrangement of network chain junctions in filled systems [60]. A model of stress-induced rupture or separation of network chains from the filler surface has been derived by Govindjee and Simo [69], who developed a complete macroscopic constitutive theory on the basis of statistical mechanics. A remarkable approach has been proposed by Witten et al. [21], who found a scaling law for the stress-strain behavior in the first stretching cycle by modeling the breakdown of a CCA-network of filler particles. They used purely geometrical arguments by referring to the available space for the filler clusters in strained samples, leading to universal scaling exponents that involve the characteristic fractal exponents of CCA-clusters. However, they did not consider effects coming in from the rubber matrix or the polymer-filler interaction strength, though these are evident from experimental data, e.g., the impact of matrix cross-linking or filler surface treatment (graphitization) on stress-strain curves. It indicates that stress-induced breakdown of filler clusters takes place, where the stress on the filler clusters is transmitted by the rubber matrix.

The above interpretations of the Mullins effect of stress softening ignore the important results of Haarwood et al. [73, 74], who showed that a plot of stress in second extension vs ratio between strain and pre-strain of natural rubber filled with a variety of carbon blacks yields a single master curve [60, 73]. This demonstrates that stress softening is related to hydrodynamic strain amplification due to the presence of the filler. Based on this observation a micro-mechanical model of stress softening has been developed by referring to hydrodynamic reinforcement of the rubber matrix by rigid filler clusters that are irreversibly broken during the first deformation cycle [75, 76]. Thereby, a non-Gaussian tube model of rubber elasticity has been applied [24, 62, 77, 78].

In the present chapter we will first focus on the specific morphology and energetic surface properties of carbon black (Sect. 3). In particular, we will consider the surface roughness and activity on atomic length scales as obtained by recent gas adsorption investigations [20, 79–82]. Then, we will concentrate on the primary aggregate structure that can be well characterized by TEM-measurements [83–85]. Both morphological characteristics of carbon black are compared to results of computer simulations and analytical models of disorderly growth processes under ballistic conditions [18, 84]. In Sect. 4, we will consider the structures involved in secondary aggregation and networking of carbon black in elastomers [22, 23] that has also been extended to elastomer blends [86, 87]. We will discuss the developed model in the framework of experimental results concerning filler flocculation and dielectric properties of carbon black reinforced rubbers. Based on the investigations of carbon black morphology and energetic surface structure we will then develop a model of rubber reinforcement by kinetically aggregated filler networks in Sect. 5. We will first consider the linear viscoelastic behavior by referring to the elasticity of the fractal CCA-unit cells of the filler network [22, 23, 48, 57, 85, 88]. Then, we will focus on a micro-mechanical concept of stress-induced filler cluster breakdown that combines the hyperelastic response of the hydrodynamically reinforced rubber matrix with the non-linear viscoelastic response of reversibly broken filler clusters [75, 76, 89]. It allows for an explanation of both, the Payne effect and the Mullins effect, by referring to a single micro-mechanical mechanism.

2 Experimental Methods

2.1 Materials and Sample Preparation

For the preparation of filler reinforced elastomer composites, most frequently commercial rubber grades with variable microstructure and broad molar mass distribution are applied. The typical rubber grades, considered in the present review, are as follows:

1. Natural rubber NR (SMR 5L; 99.9 vol.% *cis*)
2. Solution-styrene-butadiene rubber S-SBR (Buna VSL-2525-0; 25 vol.% vinyl, 25 vol.% styrene, M_w=206,600 g/mol, M_w/M_n=2.75)
3. Emulsion-styrene-butadiene rubber E-SBR (Intol 1524; 15 vol.% vinyl, 23.5 vol.% styrene, M_w=205,300 g/mol, M_w/M_n=2.12)
4. Ethylene-propylene-diene rubber EPDM (Keltan 512; 55 vol.% ethylene)
5. Nitrile-butadiene rubber NBR (Perbunan 3307; 33 vol.% acryl-nitrile)

In addition to the commercial rubber grades, model polymers with variable molar mass and narrow molecular weight distribution are often used for studying reinforcement mechanisms. (In Sect. 4.3 we consider a special S-SBR type with 29 vol.% vinyl and 25 vol.% styrene units; M_w/M_n=1.1).

In general, furnace carbon blacks with different structure and specific surface area are employed as reinforcing fillers. Furthermore, commercial silica grades are applied, especially in the tire industry. In Sect. 5 we refer to a highly dispersive silica grade (Ultrasil 7000 GR) together with a bi-functional silane as coupling agent (Si 69).) The filler loading can range from very small amounts up to 100 phr. The notation "per hundred rubber" (phr), representing the mass (in grams) of an ingredient with respect to 100 g of rubber, is widely used in the rubber industry. The conversion to volume fraction Φ is obtained by the elementary formula $\Phi = \mathrm{phr}/\rho_{CB}\,(100\ \mathrm{g}/\rho_{Pol} + \mathrm{phr}/\rho_{CB})^{-1}$, where phr is the mass of carbon black (in grams) per 100 g polymer, while $\rho_{Pol} \approx 0.9$ g/cm^3 and $\rho_{CB} \approx 1.8$ g/cm^3 is the mass density of the pure polymer and carbon black, respectively.

As polymeric model fillers with specific surface groups and narrow size distribution, different microgel types, e.g., poly(styrene)-microgel, PS(m), and poly(methoxy-styrene)-microgel, PMS(m), are applied. They can be prepared by emulsion polymerization techniques as described, e.g., in [48].

The rubber composites are in general processed in an interlocking mixer, as widely used in rubber industry, e.g., Werner & Pfleiderer GK 1.5 E with a mixing chamber of 1500 cm^3. For obtaining a sufficient dispersion of the filler, at least 4 min mixing time are necessary. For the silica/silane systems, most frequently a two step mixing procedure is applied, for obtaining a reasonable coating reaction of the silane on the silica surface. Furthermore, a small amount of ZnO and stearinic acid is added to all compounds for supporting a fast filler dispersion and vulcanization. In general, the mixer is filled up to 75% of its capacity for the specimens with less than 40 phr filler, 70% for loading between 40 and 70 phr, and 60% for more than 70 phr filler. The cross-linking system is added on a roller mill in a separate mixing step (Here, 1.7 phr sulfur and 2.5 phr *N*-cyclohexylbenzothiazol-2-sulfenamide (CBS) is used). The composites with graphitized blacks are mixed on a roller mill for 20 min to ensure a sufficient dispersion. The composites are cured in a steam press, most frequently up to 90% of the rheometer optimum (T_{90} time) at 160 °C.

2.2 Gas Adsorption Measurements

The morphological and energetic surface characterization of carbon black on atomic length scales can be performed by volumetric gas adsorption techniques. A schematic view of three different application regimes of this technique is shown in Fig. 4. According to this scheme, the surface roughness of particular fillers is estimated in the mono- and multi-layer regime. Furthermore, a characterization of the energy distribution of adsorption sites of carbon black and other colloids is obtained in the sub-layer regime.

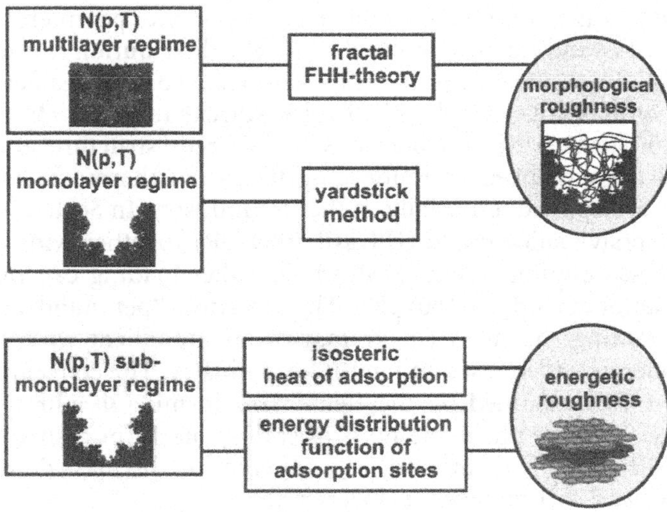

Fig. 4 Schematic representation of gas adsorption techniques in different layer regimes (decreasing pressure from *top to bottom*)

A classical volumetric adsorption apparatus equipped with absolute capacitance pressures transducers can be used for the estimation of adsorption isotherms in the pressure range 10^{-3} mbar$<p<10^{3}$ mbar. Before adsorption measurements the carbon black samples are extracted with toluene and water/methanol (1:1) and after drying degassed overnight at 200 °C at a pressure below 10^{-4} mbar. The time allowed for equilibrium of each point of the isotherm is 5–90 min depending on the sample and the adsorbed amount.

2.3 Microscopic Techniques

The macro-dispersion of the fillers can be determined by light-microscopic techniques with computer-assisted image processing on glazed cuttings of the vulcanized samples. At least five picture details have to be evaluated for each specimen. The dispersion coefficient D is calculated from the ratio of non-dispersed filler agglomerates and the volume fracture Φ of the filler in the composites in accordance with ASTM:D2663.

For the characterization of micro-dispersion of fillers, i.e., in rubber primary aggregate morphology of carbon black, the uncured composites are immersed for a week in a good solvent, with the solvent being changed a number of times in order to remove the unbounded polymer. Afterwards the specimens are dispersed in a vibrator. The highly diluted suspensions are then dipped on a grid and carefully condensed. Micrographs are taken using the Electron-Spectroscopy-Imaging-Transmission-Electron-Microscopy (ESI-TEM) technique, e.g., on an EM 902 (Zeiß) equipment. For the evaluation of aggregate morphology (analogous to ASTM: 3849) roughly 500 particles of each carbon black type are measured with respect to cross-sectional area A, perimeter P and diameter d. This preparation procedure is indicated by the terminology "in rubber state".

In a second preparation procedure, referred to as "dry state", the carbon black is resolved in a solvent and dispersed by ultrasonic treatment. Afterwards, the highly diluted suspension is dipped on a grid and treated as above.

Further morphological investigations by TEM are performed on ultra-thin cuts (~80 nm). By making use of energy filters (ESI), the inelastically scattered electrons can be removed. The micrographs provided by the elastically scattered electrons only show a good contrast based on the differences in carbon densities in the phases under consideration.

2.4 Mechanical Analysis

Dynamic-mechanical testing of cross-linked samples are often carried out with high precision on specimen strips in torsion mode, e.g., with a Rheometrics Dynamic Analyzer II (RDA) with a sample size of 28×10×2 mm. Here, temperature-and strain sweeps are performed in a displacement range from 0.01% to about 5% strain and a frequency range between 0.1 and 100 Hz. Dynamic mechanical testing of uncross-linked samples can be made, e.g., with a Rubber Process Analyzer RPA 2000 (Alpha Technologies) from 0.28% to 350% strain at various frequencies and elevated temperatures.

The quasistatic mechanical testing of the vulcanized samples is performed with a tensile tester. Uniaxial stress-strain measurements are carried out on (S2) strip-samples with cyclically increasing load (discontinuous damage mode) and up to rupture stress. Equi-biaxial stress-strain measurements are performed on a special frame with quadratic sample sheets of size 100×100 mm and thickness 2 mm up to 110% strain maximum. During the tests, the stretching velocity is chosen to be small in order to avoid dynamic contributions to the moduli. In general, 10 mm/min is used, corresponding to a strain rate of $\partial\varepsilon/\partial t \approx 4\times10^{-3}\,s^{-1}$.

2.5 Dielectric Measurements

For dielectrical investigations, samples are prepared as cross-linked sheets of a thickness between 1 and 2 mm. The dielectric measurements in a frequency range from 10^{-1} to 10^7 Hz are performed using a frequency response analysis system, e.g., the computer controlled Solartron SI 1260 Impedance/Gain-Phase Analyzer and a Novocontrol broadband dielectric converter. The high frequency measurements in the range from 10^6 to 10^9 Hz can be performed in the wave reflection mode, e.g., with an Agilent 4291B RF Impedance Analyzer.

3 The Disordered Structure of Carbon Black

Reinforcement of elastomers by colloidal fillers, like carbon black or silica, plays an important role in the improvement of mechanical properties of high performance rubber materials. The reinforcing potential is mainly attributed to two effects:

1. The formation of a physically bonded flexible filler network
2. Strong polymer filler couplings

Both of these effects refer to a high surface activity and specific surface of the filler particles [26, 27, 47]. In view of a deeper understanding of such structure-property relationships of filled rubbers it is useful to consider the morphological and energetic surface structure of carbon black particles as well as the primary and secondary aggregate structure in rubber more closely.

3.1 Surface Roughness and Activity on Atomic Length Scales

3.1.1 *Universality of Carbon Black Surface Roughness*

For the characterization of surface roughness of carbon blacks, different experimental techniques have been applied in the past. Beside microscopic investigations, e.g., AFM that give an impressive but more qualitative picture [90–92], scattering techniques such as SANS [93] and SAXS [94–96] as well as gas adsorption techniques [79–84, 97–99] have been used for a fractal analysis of surface roughness. The results discussed in the literature appear somewhat controversy, since almost flat surfaces with $d_s \approx 2$ [98, 99] and also rough surfaces with $2.2<d_s<2.6$ [79–84, 93–96] are found.

The reason for these discrepancies lies on the one hand in the restricted resolution of SANS and SAXS, since the scattering data could only be evaluated for wave vector $q_s<1\ \mathrm{nm}^{-1}$ in most cases. This corresponds to length scales larger than about 6 nm, while the gas adsorption data typically were obtained at length scales smaller than 6 nm. Recent investigations by SAXS have been extended down to smaller length scales with $q_s>1\ \mathrm{nm}^{-1}$, where a scattering from the graphitic layers at the carbon surface was observed. It means that the surface scattering was shielded by that of sheet like structures [94]. On the other hand the discrepancies between the gas adsorption results arise primary from the evaluation procedure of the effective cross-sections σ of the different gases, as far as the yardstick method in the monolayer regime is concerned. The estimation of surface fractal dimensions in the multi-layer regime is complicated by the fact that contributions of two different surface potentials have to be considered, resulting from van-der-

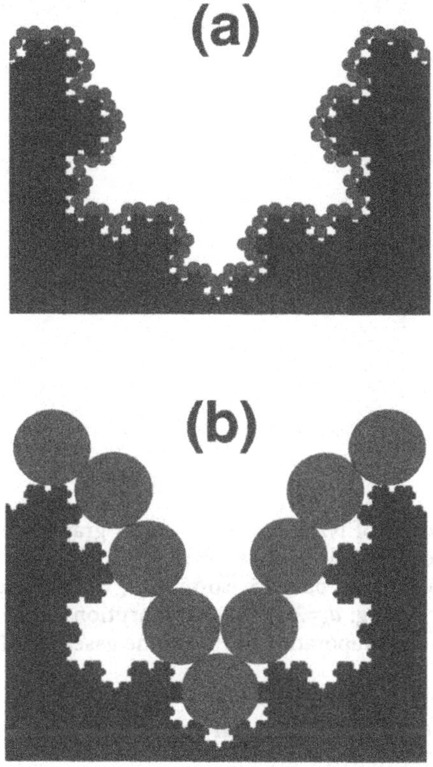

Fig. 5a,b Schematic presentation of a mono-layer coverage of a fractal surface with: **a** small; **b** large gas molecules

Waals- and surface tension interaction, respectively. Dependent on the dominance of one of the two potentials, remarkably different estimates of surface roughness are obtained. For that reason a proper analysis of these factors is necessary for getting reliable results.

In the present chapter we will summarize results of two different evaluation procedures for the surface roughness of carbon blacks. In the mono-layer regime we refer to the scaling behavior of the estimated BET-surface area with the size of adsorbed probe molecules (yardstick method). On smooth flat surfaces the BET-area is independent of the adsorbed probe or applied yardstick, while on rough surfaces it decreases with increasing probe (yardstick) size due to the inability of the large molecules to explore smaller cavities. This is shown schematically in Fig. 5.

In the case of carbon black a power law behavior of the BET-surface area with varying yardstick size is observed, indicating a self-similar structure of the carbon black surface. Double logarithmic "yardstick-plots" of the BET-mono-layer amount N_m vs cross-section σ of the probe molecules are shown in Fig. 6 for an original furnace black N220 and a graphitized (T=2500 °C) sample N220g. It demonstrates that the roughness exponent or surface frac-

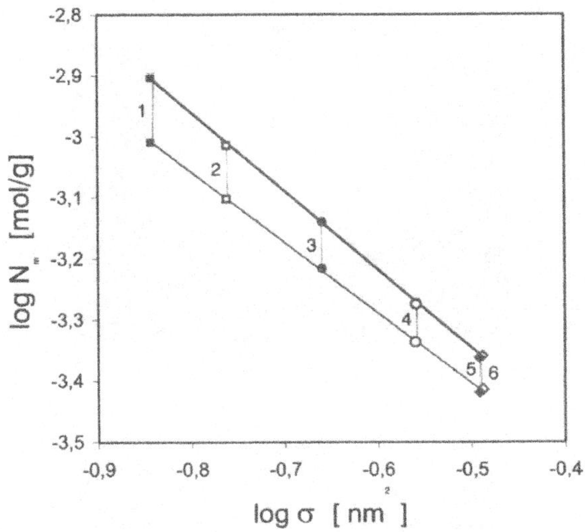

Fig. 6 Yardstick-plot (Eq. 1) of N220 (*triangle*) and a graphitized N220g (*filled circles*) with adsorption cross section σ determined from the bulk liquid density ρ (Eq. 2); 1 argon, 2 methane, 3 ethane, 4 propane, 5 iso-butane, 6 *n*-butane; The slopes yield for N220: d_s=2.56±0.04, for N220g: d_s=2.32±0.03. Adsorption temperatures and densities ρ are chosen according to the evaporation points of the gases at 1000 mbar

tal dimension d_s differs for the two carbon black samples. By using the relation introduced by Mandelbrot [1, 2]:

$$N_m \sim \sigma^{-\frac{d_s}{2}} \tag{1}$$

one obtains from the slopes of the two regression lines of Fig. 6a surface fractal dimension $d_s \approx 2.56$ for the N220-sample and $d_s \approx 2.32$ for the graphitized N220g sample. An extrapolation of both regression lines yields an intersection at an ultimate cross-section that corresponds to a yardstick length of about 1 nm, indicating that graphitization reduces the roughness of carbon black on small length scales below 1 nm, only. Figure 6 clearly demonstrates that the reduction of BET-surface area during graphitization is length scale (yardstick) dependent, proving that it is related to a change of surface morphology and not, e.g., a result of reduced energetic surface activity (see below).

An important point in the above evaluation of carbon black surface morphology is the correct estimation of the cross-section σ of the applied probe molecules. This is done by referring to the mass density ρ of the probe molecules in the bulk liquid state that are considered as spheres in a hexagonal close packing:

$$\sigma = 1.091 \left(\frac{M}{N_a \rho}\right)^{2/3} \tag{2}$$

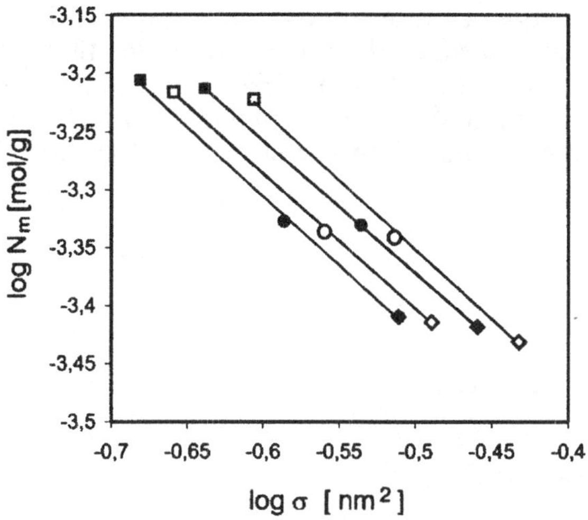

Fig. 7 Yardstick-plots (Eq. 1) of the graphitized black N220g obtained with a series of alkenes (ethylene, propylene, iso-butylene) (*filled symbols*) and alkanes (ethane, propane, iso-butane) (*open symbols*). Adsorption temperatures are chosen as evaporation points at vapor pressures $p_0 \approx 10^3$ mbar (*lower curves*) and $p_0 \approx 10^4$ mbar (*upper curves*) of the condensed gases, respectively

Here, M is the molar mass of the probe molecules and N_a is the Avogadro number. The crucial point now is the temperature dependence of ρ that differs for the different probe molecules, mainly due to variations in the characteristic temperatures, e.g., the evaporation points.

We found that Eq. (2) can be applied without further corrections and high correlation coefficients of the "yardstick-plots" in Fig. 6 and Fig. 7 are obtained only if the temperature during the adsorption experiments of a chemically similar, homological series of gases is chosen according to the same reference state, as defined in the framework of the theory of corresponding states. This is demonstrated in the "yardstick-plots" of Fig. 7, showing that for the same carbon black (N220g) a different scaling factor is obtained within one series of gases, if the adsorption temperatures are chosen with respect to different reference pressures, i.e., the evaporation temperatures at $p_o=10^3$ mbar and $p_o=10^4$ mbar, respectively.

As shown in Fig. 7, a different scaling factor is also observed for the two different homological series of gases, i.e., the alkanes and alkenes, respectively. However, the scaling exponent and hence the surface fractal dimension $d_s \approx 2.3$ is unaffected by the choice of the reference pressure or applied series of adsorption gases.

An alternative approach to the characterization of surface morphology of carbon blacks is the consideration of film formation of adsorbed molecules in the multi-layer regime. In this case, the surface roughness is evaluated with respect to a fractal extension of the classical Frenkel-Halsey-Hill (FHH)-theory, where, beside the van der Waals surface potential, the vapor-liquid surface tension has to be taken into account [100, 101]. Then the

Helmholtz free energy of the adsorbed film is given as the sum of the van der Waals attraction potential of all molecules in the film with all atoms in the adsorbent, the vapor-liquid surface free energy and the free energy of all molecules in the bulk liquid. This leads to the following relation between the adsorbed amount N and the relative pressure p/p_0 [100, 101]:

$$N \sim \left(\ln \frac{p_o}{p} \right)^{-\vartheta} \tag{3}$$

with

$$\vartheta = \frac{3 - d_s}{3} \quad \text{FHH} - \text{regime} \tag{3a}$$

$$\vartheta = 3 - d_s \quad \text{CC} - \text{regime} \tag{3b}$$

The different exponents for the FHH- and capillary condensation (CC)-regime consider the two cases where adsorption is dominated by the van der Waals potential and the vapor-liquid surface tension, respectively. The two cases are shown schematically in Fig. 8b,c, respectively. Note that in the CC-regime a flat vapor-liquid surface is obtained due to a minimization of curvature by the surface tension. In contrast, in the FHH-regime the vapor-liquid surface is curved, since it is located on equi-potential lines of the van der Waals potential with constant distance to the adsorbent surface.

At low relative pressures p/p_0 or thin adsorbate films, adsorption is expected to be dominated by the van der Waals attraction of the adsorbed molecules by the solid that falls off with the third power of the distance to the surface (FHH-regime, Eq. 3a). At higher relative pressures p/p_0 or thick adsorbate films, the adsorbed amount N is expected to be determined by the surface tension γ of the adsorbate vapor interface (CC-regime, Eq. 3b), because the corresponding surface potential falls off less rapidly with the first power of the distance to the surface, only. The cross-over length $z_{\text{crit.}}$ between both regimes depends on the number density n_{p} of probe molecules in the liquid, the surface tension γ, the van der Waals interaction parameter α as well as on the surface fractal dimension d_s [100, 101]:

$$z_{crit} = \sqrt{\frac{\alpha n_p}{(d_s - 2)\gamma}} \tag{4}$$

Note that the cross-over length $z_{\text{crit.}}$ decreases with increasing surface fractal dimension d_s, implying that the FHH-regime may not be observed on very rough surfaces, i.e., the film formation may be governed by the surface tension γ on all length scales $z>a$ (compare Fig. 10).

The film thickness z is related to the surface relative coverage N/N_{m} and the mean thickness $a\approx0.35$ nm of one layer of nitrogen molecules [102] according to the scaling law [1, 2]:

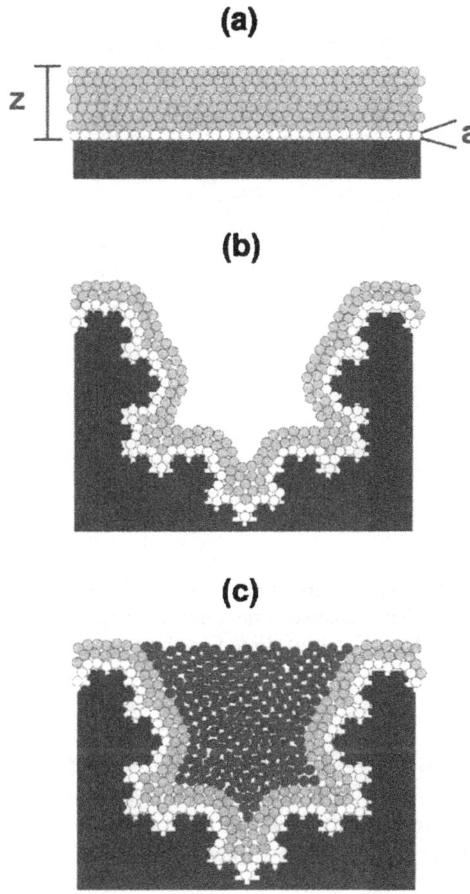

Fig. 8a–c Schematic view of the coverage of: **a** a smooth; **b,c** a fractal surface according to the fractal FHH-theory: (*open circles*) monolayer-regime, (*gray circles*) FHH-regime, (*filled circles*) CC-regime; *z*: average film thickness, *a*: monolayer thickness

$$\frac{N}{N_m} = \left(\frac{z}{a}\right)^{3-d_s} \tag{5}$$

The monolayer amount N_m can be estimated from a classical BET-plot and hence the film thickness z can be obtained directly from the adsorbed amount N if the surface fractal dimension d_s is known.

So-called FHH-plots of the nitrogen adsorption isotherms at 77 K of various graphitized furnace blacks are shown in Fig. 9. The graphitized furnace blacks have two linear ranges. Starting from low pressures (right side), the first linear range is fitted by Eq. (3a), because the film is not very thick and the van der Waals attraction of the molecules by the solid governs the adsorption process (FHH-regime). With rising pressure, at a critical film thickness of about $z_{crit} \approx 0.5$ nm (Eq. 5), the vapor-liquid surface tension γ becomes dominant and a step-like increase of the adsorbed amount is observed. The fractal FHH-theory claims fractal dimensions of $d_s \approx 2.3$ up to a

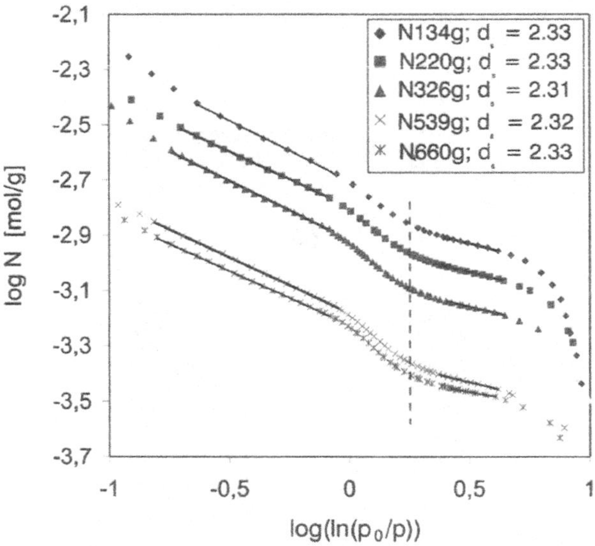

Fig. 9 FHH-plot of nitrogen adsorption isotherms at 77 K on various graphitized furnace blacks, as indicated. The *dashed line* characterizes the transition between the FHH- and the CC-regime. The d_s-values, listed in the insert, refer to the FHH-regime at low pressures

length scale of $z \approx 1$ nm, independent of grade number. At this length scale a geometrical cut-off appears and the surface becomes rougher. In the final linear regime, corresponding to $z > 1$ nm, the surface fractal dimension takes the value $d_s \approx 2.6$ (CC-regime). This linear range has an upper cut off length of $z \approx 6$ nm.

Figure 10 shows that contrary to the graphitized blacks the untreated furnace blacks have only one linear range with a fractal dimension of $d_s \approx 2.6$ (CC-regime, Eq. 3b). Obviously the van der Waals attraction can be neglected and the surface tension γ controls the adsorption process on all length scales. This is due to the larger surface fractal dimension d_s as compared to the graphitized furnace blacks that shifts the cross-over length $z_{crit.}$ to smaller values (Eq. 4). Assuming that the number density n_p, the surface tension γ of the adsorbate and the van der Waals interaction parameter α are approximately the same for liquid nitrogen adsorbed on graphitized and untreated furnace blacks, a cross-over length of $z_{crit} \approx 0.35$ nm can be estimated from Eq. (4) with the experimental values of the fractal dimensions and the crossover length $z_{crit.} \approx 0.5$ nm on a graphitized carbon black. The value $z_{crit.} \approx 0.35$ nm is already in the range of the detection limit given by the layer thickness $a \approx 0.35$ nm. Hence, the nitrogen adsorption on furnace carbon blacks is dominated by the vapor-liquid surface tension on all length scales and a cross-over between the FHH- and the CC-regime does not appear.

The results for the surface fractal dimension of a series of furnace blacks and graphitized blacks, obtained by nitrogen adsorption in the multilayer regime, are summarized in Fig. 11. The cut off lengths are quite similar

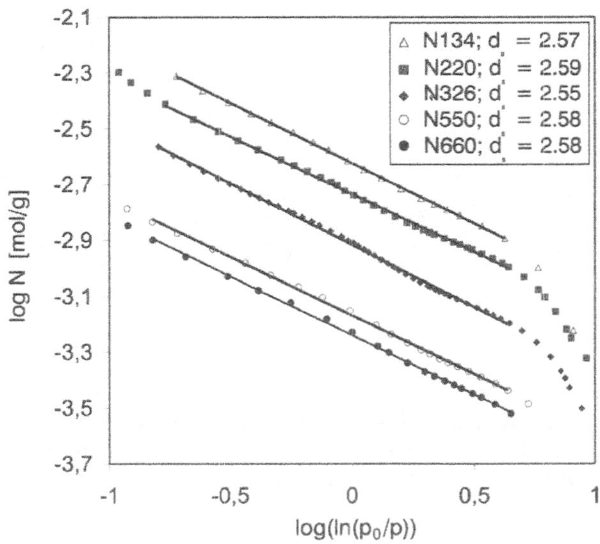

Fig. 10 FHH-plots of nitrogen adsorption isotherms at 77 K of various furnace blacks, as indicated. The surface fractal dimension appears to be universal, i.e., it varies between d_s=2.55 and d_s=2.59 for the depicted furnace blacks

within both series of blacks and agree with those found in Fig. 9 and Fig. 10. In Fig. 11 a weak trend of increasing surface fractal dimension with increasing specific surface (decreasing primary particle size) is observed. This results from increasing curvature of the particle surface with decreasing size, since crystallite structures with edges are present on the surface that lead to a more pronounced roughness, if arranged on a strongly curved surface. We will see in the next section that the amount of crystallite edges and slit shaped cavities increases slightly with increasing specific surface, leading to a more pronounced energetic surface activity for the fine carbon blacks. This relatively small effect correlates well with the weak trend of the surface fractal dimension observed in Fig. 11.

The observed almost universal value of the surface fractal dimension $d_s \approx 2.6$ of furnace blacks can be traced back to the conditions of disordered surface growth during carbon black processing. It compares very well to the results evaluated within the an-isotropic KPZ-model as well as numerical simulations of surface growth found for random deposition with surface relaxation. This is demonstrated in some detail in [18].

3.1.2
Energy Distribution of Carbon Black Surfaces

The energy distribution $f(Q)$ of carbon black surfaces is calculated by assuming that the measured overall isotherm consists of a sum of generalized Langmuir isotherms of various interaction energies Q, implying that the energy distribution can be identified with the numerically obtained weighting

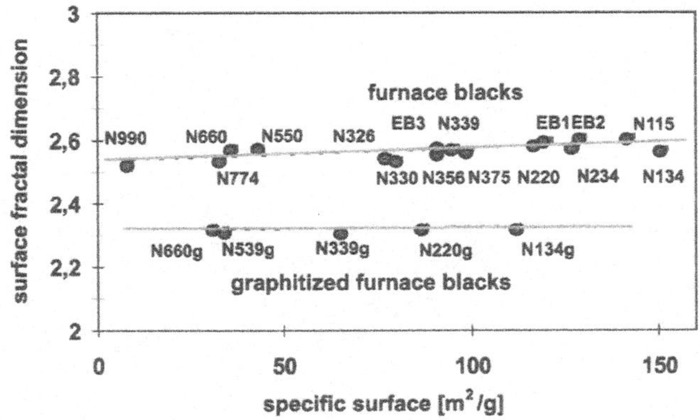

Fig. 11 Surface fractal dimensions d_s on atomic length scales of furnace blacks and graphitized blacks in dependence of specific surface. The data are obtained from nitrogen adsorption isotherms in the multilayer regime

function [103, 104]. For a continuous distribution function $f(Q)$ the overall isotherm $\Theta(p,T)$ is given by

$$\Theta(p,T) = \int_0^\infty \theta(p,T,Q) \cdot f(Q)\, dQ \tag{6}$$

The integral in Eq. (6) is normalized to unity. This has to be taken into consideration if solid samples with different specific surface areas are compared.

For an evaluation of the local model isotherm $\theta(p,T,Q)$ with constant interaction energy Q, the effects of multi-layer adsorption and lateral interactions between neighboring adsorbed molecules are considered by applying two modifications to the Langmuir isotherm: (i) a multi-layer correction according to the well known BET-concept and (ii) a correction due to lateral interactions with neighboring gas molecules introduced by Fowler and Guggenheim (FG) [105]:

$$\theta(p,T,Q) = \frac{b_{BET}^2 \cdot b_{FG} \cdot b_L \cdot p}{1 + b_{BET} \cdot b_{FG} \cdot b_L \cdot p} \tag{7}$$

with

$$b_{BET} = \frac{1}{1 - \frac{p}{p_0}} \tag{8}$$

$$b_{FG} = e^{\frac{z_n \chi \theta}{RT}} \tag{9}$$

$$b_L = \frac{N_a \sigma \tau_0}{\sqrt{2\pi MRT}} \cdot e^{\frac{Q}{RT}} \tag{10}$$

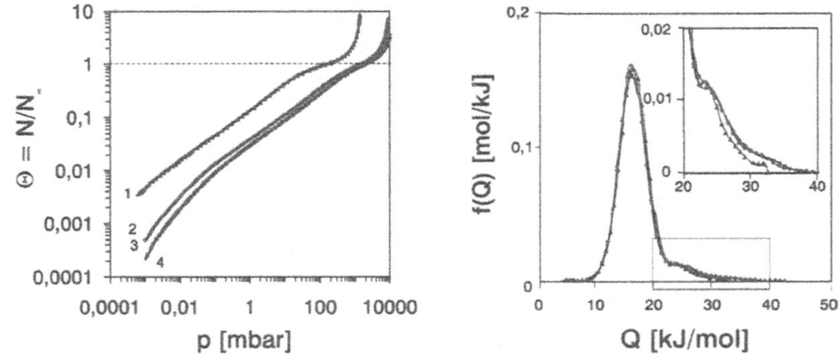

Fig. 12 Adsorption isotherms ($\Theta=N/N_m$) and corresponding energy distribution functions of ethylene on N220 at various temperatures; ((1) T=177 K; (2),(3) T=223 K; (4) T=233 K)

Here, z_n is the number of neighboring adsorption sites, χ is the contribution of the lateral interaction to Q, and θ is the probability that the neighboring sites are occupied by a gas molecule. R is the gas constant, T is temperature, τ_0 is Frenkel's characteristic adsorption time, σ is the adsorption cross-section of the gas molecules, N_a is Avogadro's constant, and M is molar mass. Note that the probability that the neighbor sites are occupied with other gas molecules is taken to be the local surface coverage θ and not the overall surface coverage Θ. This means that sites with the same interaction energies Q are assumed to be arranged in patches, which is in accordance with the picture of graphite like micro crystallites on the surface of the carbon black particles [26]. A probability Θ stands for a random distribution of sites with Q.

By referring to adsorption isotherms of ethylene down to very low surface coverings (10^{-3} to 1 mono-layers), the energy distribution function of adsorption sites on different furnace blacks was estimated with Eqs. (6) to (10) by applying a numerical iteration procedure. This is described in some detail in [20, 79]. For a test of the evaluation procedure, the resulting energy distribution functions obtained from four different isotherms (three different temperatures) of ethylene on N220 are compared in Fig. 12. It becomes obvious that the isotherms measured at different temperatures lead to the same result for the energy distribution function, approximately, confirming the applied procedure.

An analysis of the energy distribution function of ethylene on N220 is shown in Fig. 13, where the distribution function is fitted to four different Gauss-functions. Obviously, the good fit indicates that four different types of adsorption sites can be distinguished on the N220 surface. We relate the low energetic peak (I) to the basaltic layers and peak (III) to the edges of carbon crystallites. Peak (II) is referred to amorphous carbon and peak (IV) results from a few highly energetic slit like cavities between carbon crystallites. This is shown schematically in Fig. 14.

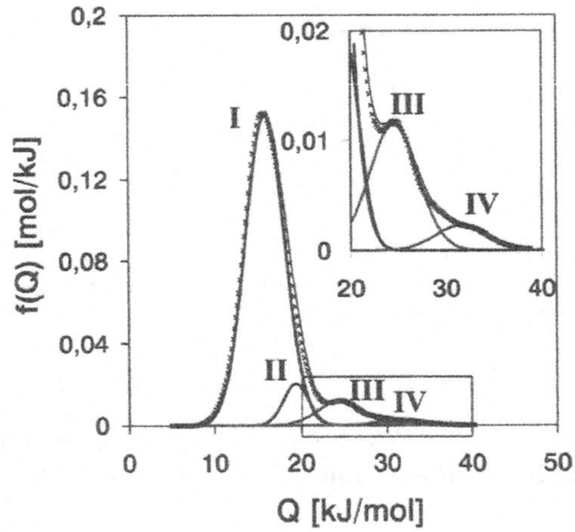

Fig. 13 Fitting of the energy distribution function of ethylene on N220, already shown in Fig. 12 (T=223 K), to four Gaussian-peaks (I–IV)

The attachment of peak (II) to the amorphous carbon is concluded from the observation that this peak does not appear in the case of the graphitized black N220g and graphitic powder. The corresponding isotherms and energy distribution functions are depicted in Fig. 15. A comparison to the above analyzed N220 and a strongly reinforcing channel gas black demonstrates the relatively large amount of highly energetic sites of these blacks (Fig. 15).

A comparison of the adsorption isotherms and the resulting energy distribution functions of three different furnace blacks is shown in Fig. 16. Corresponding to the difference in level and shape of the isotherms the amount of highly energetic sites varies significantly. The black N115 has a large fraction of highly energetic sites, while the N550 shows a small fraction of highly en-

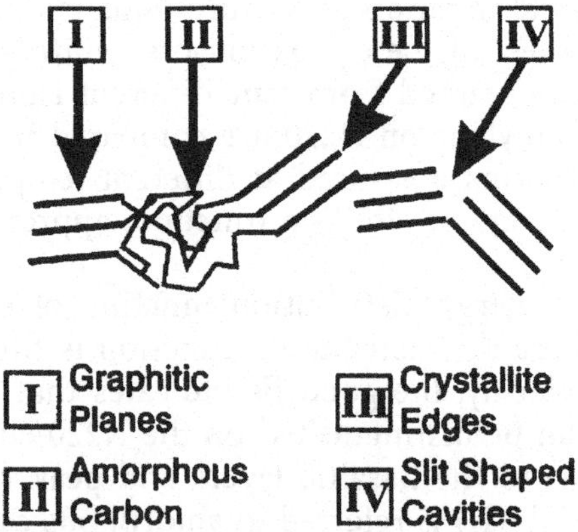

Fig. 14 Schematic view of the association between morphological arrangements of carbon crystallites and energetic characteristics of carbon black surfaces. Four different types of adsorption sites are distinguished that refer to the de-convolution shown in Fig. 13

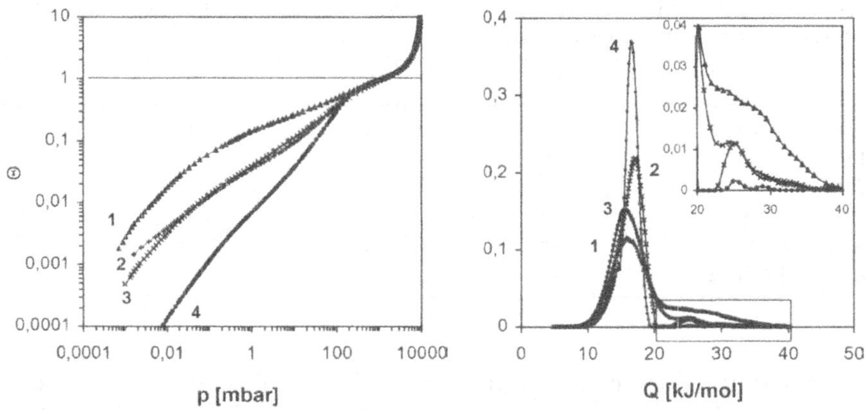

Fig. 15 Adsorption isotherms and evaluated energy distribution functions of ethylene on four different colloidal fillers at T=223 K; (1) channel gas black; (2) graphitic powder; (3) N220; (4) graphitized N220g

ergetic sites. The results of the peak analysis for the examined blacks and the graphitic powder are quantified in Table 1. A more detailed representation of the energetic surface heterogeneity of carbon blacks is found in [20, 79].

Summarizing we can conclude from the analysis of the surface energy distribution of carbon blacks that four different energetic sites can be distinguished. The fraction of highly energetic sites decreases significantly with grade number and disappear almost completely during graphitization. It indicates that the reinforcing potential of carbon black is closely related to the amount of highly energetic sites that can be well quantified by the applied gas adsorption technique. Theoretical investigations on the effect of morphological as well as energetic surface roughness on the polymer-filler inter-

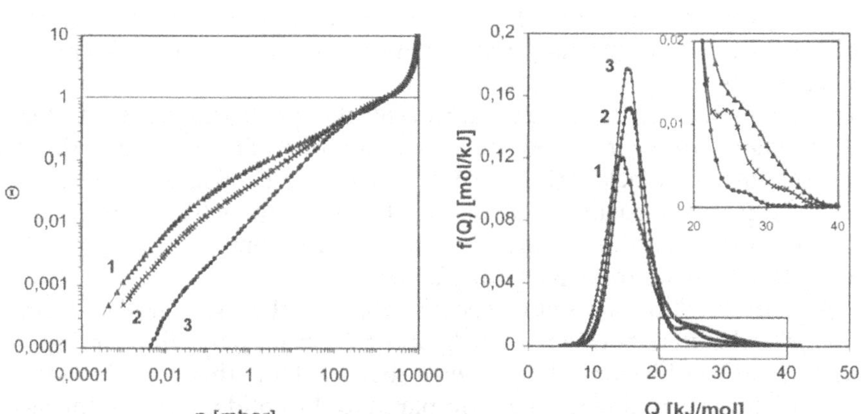

Fig. 16 Adsorption isotherms and evaluated energy distribution functions of ethylene on three different furnace blacks at T=223 K; (1) N115; (2) N220; (3) N550

Table 1 Estimated fraction [%] of the four different types of energetic sites (I–IV) for adsorption of ethylene on various colloids

	I	II	III	IV
Channel-gas-black	61	2	31	6
N115	69	13	15	3
N220	84	7	7	2
N550	93	6	1	<1
Graphitic powder	94	0	4	2
Graphitized N220g	99	0	<1	<1

action strength confirm this finding [64, 65, 106, 107]. The combination of two types of disorder, given by the pronounced morphological roughness (d_s=2.6) and the inhomogeneous energetic surface structure of carbon blacks, enhances the polymer filler coupling, significantly. It represents an important reinforcing mechanism on atomic length scales associated with the required strong phase binding in high performance elastomer composites.

3.2 Morphology of Carbon Black Aggregates on Nanoscales

Carbon blacks for the rubber industry are produced in a variety of classes and types, depending on the required performance of the final product. In general they consist of a randomly ramified composition of primary particles that are bonded together by strong sinter bridges. Significant effects of the different grades of carbon blacks in elastomer composites result from variations in the specific surface and/or "structure" of the primary aggregates [26, 27]. The specific surface depends strongly on the size of the primary particles and differs from about 10 m^2/g for the very coarse blacks up to almost 200 m^2/g for the fine blacks. The "structure" of the primary aggregates describes the amount of void volume and is measured, e.g., by oil (DBP) absorption. It typically varies between 0.3 cm^3/g and 1.7 cm^3/g for furnace blacks.

The characteristic shape of carbon black aggregates is illustrated in Fig. 17, where transmission electron micrographs (TEM) of three different grades of furnace blacks (N220, N330, N550) are shown. The variation in size of the primary particles, increasing from left to right, becomes apparent. It implies a decline of the specific surface from 116 m^2/g for N220, 81 m^2/g for N330, up to 41 m^2/g for N550.

The "structure" or amount of specific voids of the three grades is almost the same and differs between 1 cm^3/g and 1.2 cm^3/g, only. Since the specific weight of carbon black is almost twice that for DBP, this corresponds to a factor of 2 for the void volume as compared to the solid volume of the aggregates. It means that about 2/3 of the aggregate volume is empty space, i.e., the solid fraction Φ_p of the primary aggregates is relatively small ($\Phi_p \approx 0.33$).

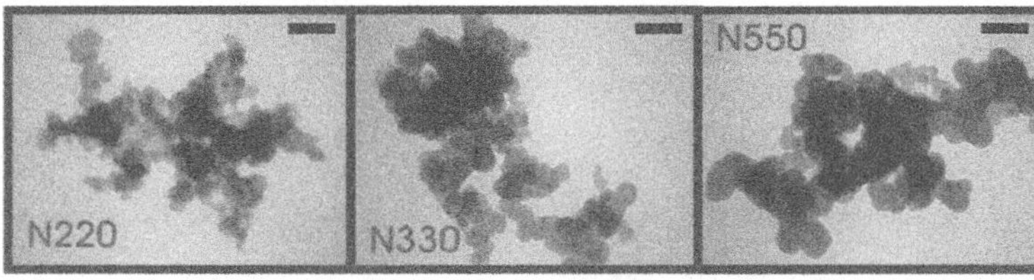

Fig. 17 Transmission electron micrographs (TEM) of three different grades of furnace blacks N220, N330, and N550 (dry state). (*Bar length*: 100 nm)

It is shown below that Φ_p fulfills a scaling relation which involves the size and mass fractal dimension of the primary aggregates. Due to significant deviations of the solid fraction Φ_p from 1, the filler volume fraction Φ of carbon black in rubber composites has to be treated as an effective one in most applications, i.e., $\Phi_{eff}=\Phi/\Phi_p$ (compare [22]).

3.2.1
Fractal Analysis of Primary Carbon Black Aggregates

For a quantitative analysis of the structure of carbon blacks as shown in Fig. 17 it is useful to consider the solid volume V_p or the number of primary particles N_p per aggregate in dependence of aggregate size d. In the case of fractal objects one expects the scaling behavior [1, 2]

$$V_p \sim N_p \sim d^{d_f} \tag{11}$$

The exponent d_f is denoted mass fractal dimension or simply fractal dimension. It characterizes the mass distribution in three dimensional space and can vary between $1<d_f<3$. This kind of fractal analysis of furnace blacks was performed, e.g., by Herd et al. [108] or Gerspacher et al. [109, 110]. The solid volume V_p of primary aggregates is normally determined (ASTM: 3849) from the cross-section area A and the perimeter P of the single carbon black aggregates by referring to a simple Euclidean relation [108]:

$$V_p = \frac{8\,A^2}{3\,P} \tag{12}$$

However, it is not quite clear whether this relation can be applied for non-Euclidean, ramified structures. Simulation results of carbon black formation under ballistic conditions by Meakin et al. [14] indicate that a scaling equation is fulfilled, approximately, between the number of particles N_p in a primary aggregate and the relative cross section area A/A_p:

$$N_p = 1.51\,(A/A_p)^{1.08} \tag{13}$$

Here, A_p is the cross section area of a single primary particle. Dependent on the application of Eqs. (12) or (13), respectively, significantly different values for the mass fractal dimension are obtained.

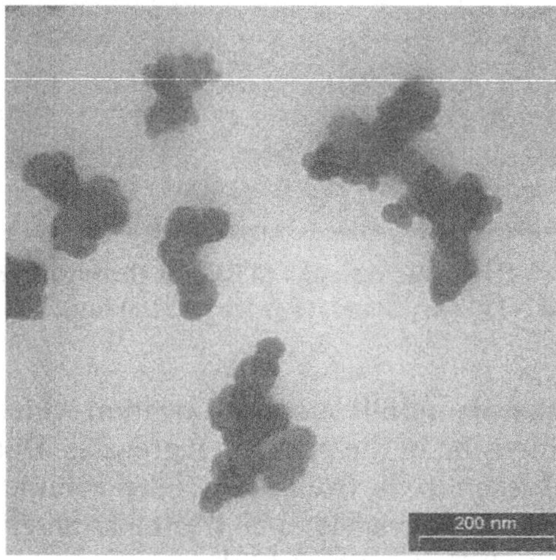

Fig. 18 TEM-micrograph of carbon black aggregates (N339) prepared from ready mixed S-SBR-composites with 60 phr filler (in-rubber state)

This discrepancy is demonstrated in Fig. 18 and Fig. 19 by considering an example of a fractal analysis of primary carbon black aggregates. Figure 18 shows a TEM-micrograph of the furnace black N339 prepared from a ready mixed composite of S-SBR after removing of the unbounded polymer as explained in Sect. 2.3 (in-rubber state). A double logarithmic plot of the solid volume V_p and the particle number N_p, estimated from Eqs. (12) and (13), vs aggregate diameter d is shown in Fig. 19a,b, respectively. The aggregate average diameter d is estimated as the mean value from 16 measurements on a single aggregate with a 15% variation in the angle of rotation. The obtained fractal dimensions differ significantly for the two evaluation procedures.

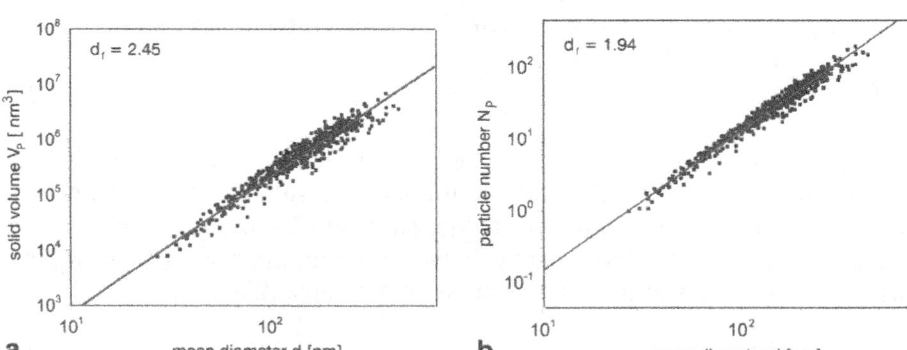

Fig. 19 **a** Fractal analysis according to Eq. (11) of primary carbon black aggregates (N339) prepared from S-SBR-composites with 60 phr filler (in-rubber state). V_p is evaluated from Eq. (12). **b** Fractal analysis according to Eq. (11) of the same set of primary carbon black aggregates (N339) as shown in Fig. 19a. N_p is evaluated by using Eq. (13)

From the slope of the two regression lines one finds d_f=2.45 and d_f=1.94, respectively.

In view of a discussion of this discrepancy, we consider the conditions of primary aggregate growth during carbon black processing in some detail. Figure 20 shows a schematic representation of carbon black formation in a furnace reactor, where a jet of gas and oil is combusted and quenched, afterwards. Beside the aggregate growth, resulting from the collision of neighboring aggregates, surface growth due to the deposition of carbon nuclei on the aggregates takes place during the formation of primary carbon black aggregates. The surface growth leads to the universal surface roughness, analyzed by gas adsorption technique in Sect. 3.1.1 and investigated from a theoretical point of view in [18]. Obviously, the surface growth is also responsible for the strength of the primary aggregates, since it proceeds in the contact range of the collided aggregates implying a strong bonding by sinter bridges (Fig. 20).

Due to the high temperature in the reactor, aggregate as well as surface growth take place under ballistic conditions, i.e., the mean free path length of both growth mechanisms is large compared to the characteristic size of the resulting structures [14–16]. Then the trajectories of colliding aggregates (or nuclei) can considered to be linear. Numerical simulations of ballistic cluster-cluster aggregation yield a mass fractal dimension d_f≈1.9–1.95 [11, 12, 17]. This compares to the above TEM-result d_f≈1.94 evaluated with Eq. (13). It means that the assumption of ballistic cluster-cluster aggregation during carbon black processing, used already in the derivation of Eq. (13), is confirmed by the TEM-data of the relatively fine black N339. For the more coarse blacks, with a typically small primary particle number, finite size effects can lead to a more compact morphology that differs from the scaling prediction of ballistic cluster aggregation. A further deviation can result from electrostatic repulsion effects due to the application of processing agents (alkali metal ions) for designing the coarse blacks (compare [22]). Note that a similar relation as Eq. (13) was derived in the 1960s by Medalia and Heckman [111, 112]. The value d_f=1.94 also agrees fairly well with other estimates obtained, e.g., by electric force microscopy [113], TEM [114], or SAXS [95, 96].

Therefore, it appears likely that the approach considering the solid volume of primary aggregates, as evaluated from the two-dimensional cross-section area by Eq. (12), leads to an overestimation of the mass fractal dimension. A more realistic estimate is obtained with Eq. (13). By referring to Eq. (12), the data obtained by Herd et al. [108] show a successively increasing value of the mass fractal dimension from d_f≈2.3 to d_f≈2.8 with increasing grade number (or particle size) of the furnace blacks. As expected, they fit quite well to the above estimate d_f≈2.45 for the black N339. A summary of these data and a discussion including other fractal parameters is found in [22].

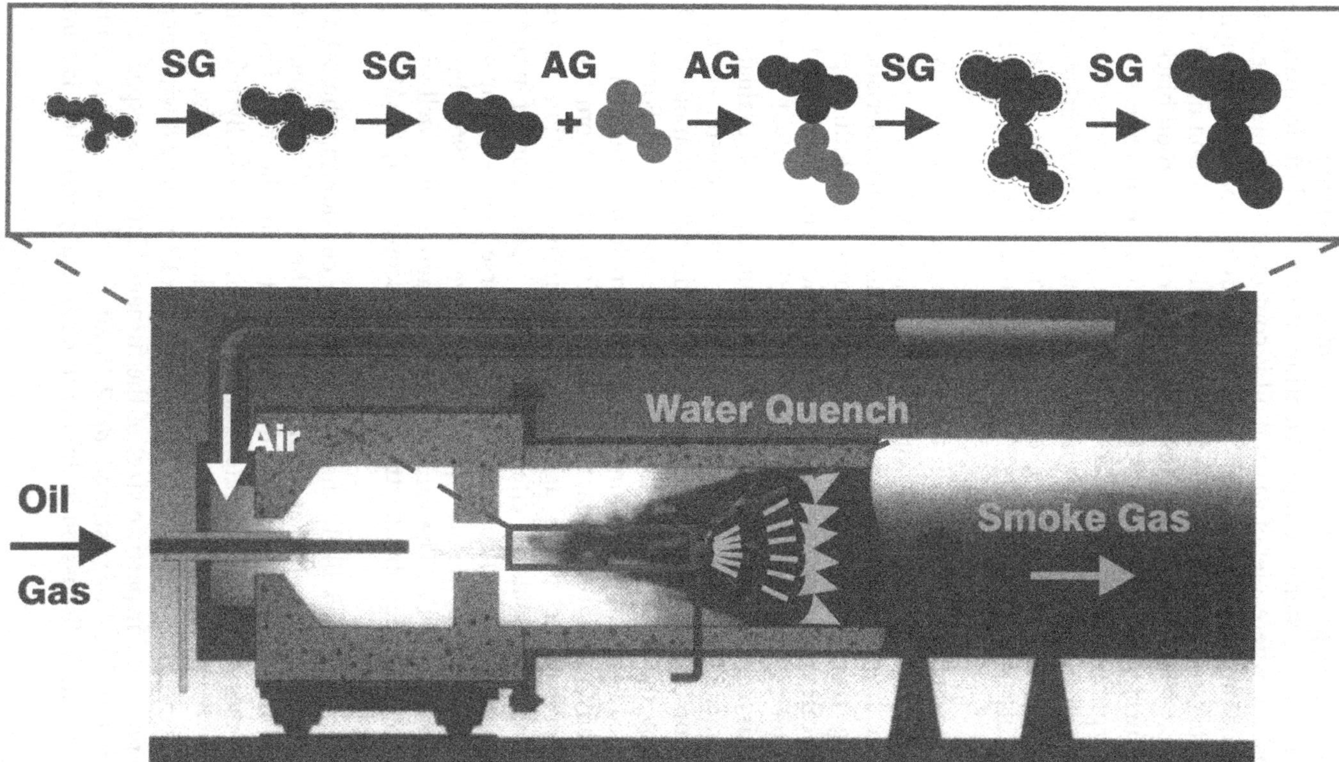

Fig. 20 Schematic view of carbon black processing in a furnace reactor. Primary aggregates are built by two simultaneous growth processes: (i) surface growth (SG) and (ii) aggregate growth (AG)

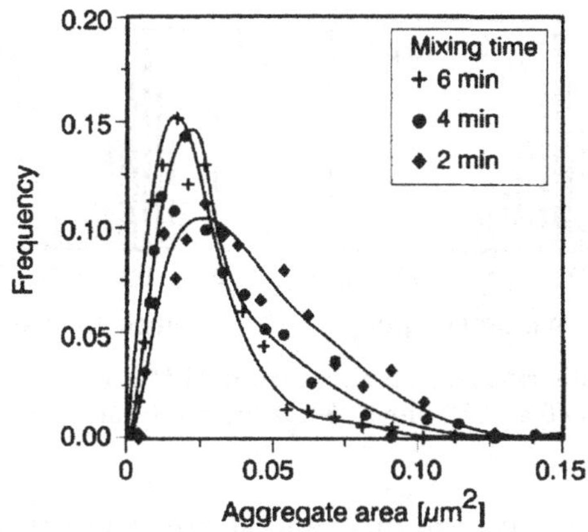

Fig. 21 Aggregate size distribution obtained from TEM-analysis (in-rubber state) of E-SBR-samples filled with 50 phr N330 at various mixing times, as indicated [115]

3.2.2
Effect of Mixing on In-Rubber Morphology of Primary Aggregates

It is well established that the specific properties of carbon black filled elastomers, e.g., viscoelasticity or electrical conductivity, are strongly affected by the disordered, ramified structure of the primary aggregates [26, 27]. On the one hand, this structure is characterized by the mass fractal dimension considered above. On the other hand, it is determined by a lower and upper cut-off length, i.e., primary particle size and aggregate size, respectively. In the following we will focus on the upper cut-off length or, more precisely, on the size distribution of primary aggregates in ready mixed composites. We will see that this quantity depends on the conditions of sample preparation, since aggregate rupture can take place if high shear stresses are applied during the mixing procedure.

In the literature it has been found that during mixing aggregate breakdown occurs for a number of carbon blacks in highly viscous rubbers [115–118]. Recently, the aggregate breakdown was also attributed to classes of specific shapes of individual carbon blacks [108]. The opinion about the mechanical consequences of this process is quite different. On the one side, no obvious relationship to reinforcement is conjectured [116]. On the other side, improvements of the mechanical performance, due to the creation of new, active carbon surface, is assumed, which participates in formation of a strong filler-rubber coupling [118].

Figure 21 shows results obtained from TEM-analysis of primary aggregate size distribution of E-SBR/N330-samples at various mixing levels. With increasing mixing time, a shift in aggregate size distribution to smaller values is observed. The maximum of the distribution of aggregate cross section area shifts from about 0.03 μm^2 to about 0.02 μm^2. The shift of the maximum can be related to a breakdown of aggregates into smaller pieces as mixing time increases. It can also be referred to an improved micro-dispersion

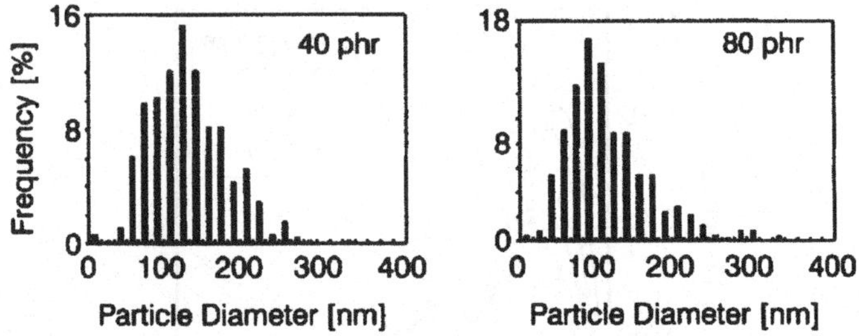

Fig. 22 Aggregate size distribution obtained from TEM-analysis (in-rubber state) of E-SBR-samples filled with 40 and 80 phr N339, respectively, at fixed mixing time of 4 min

with specific influences on the mechanical property spectrum [85]. A characteristic effect of increasing mixing time on aggregate size is the reduction in the shoulder, occurring in the case of aggregate cross section areas greater than 0.05 μm^2, in favor of smaller aggregates. This is indicative for the rupture of single aggregates into two pieces.

The observed aggregate breakdown during mixing can be understood on a more fundamental level, if the above discussed two simultaneous growth mechanisms, surface- and aggregate growth, during carbon black processing are considered again (Fig. 20). Obviously, the surface growth implies a strong bonding between adjacent primary particles by rigid sinter bridges that keep the primary aggregates together. However, this process goes on during the aggregate growth leading to a hierarchy of bonding strengths. The bonds formed in the beginning of aggregate growth become stronger than the final ones, because the time for stabilization by sinter bridges decreases with increasing time in the reaction zone. The bonds formed between collided aggregates just at the end of the reaction zone, before the quenching process takes place, remain relatively weak. This also becomes apparent in the upper scheme of Fig. 20.

For that reason we expect that increasing mixing severity during compounding with highly viscous polymer melts leads to aggregate breakdown and changes in aggregate size distributions. According to Fig. 22, this is also observed for an increased filler loading.

As shown in Fig. 22, the maximum of the size distribution is shifted to smaller values with rising filler concentration from 40 phr N339 to 80 phr N339 in E-SBR-composites. This results from the increased viscosity, since shear forces during mixing are enhanced with rising viscosity of the composite. A comparison of the morphology of N339 in E-SBR- and S-SBR-composites with increasing carbon black concentration is summarized in Table 2. It emphasizes the successive decrease of the mean primary aggregate size with increasing filler loading for both systems. The composites with E-SBR have a slightly larger aggregate size in the range of higher filler concentrations compared to those with S-SBR. On the one hand this can be assumed to result from a lower viscosity, especially under elongation deformations, which is of high relevance for filler dispersion during mixing. On the other,

Table 2 Characteristic parameters from TEM-analysis for primary aggregates of N339 in E-SBR- and S-SBR-composites, respectively, at various filler concentrations and fixed mixing time (4 min)

N339 [phr]	<*A*> [nm^2]	<*P*> [nm]	<V_P> [nm3]
E-SBR			
20	21,600	777	1,750,230
40	14,570	600	1,009,400
60	12,090	534	838,779
80	12,170	516	780,592
S-SBR			
20	22,900	857	1,711,680
40	15,820	639	1,113,475
60	12,580	539	839,255
80	8,060	399	482,050

<*A*>: mean cross section area
<*P*>: mean perimeter
<V_P>: mean solid volume (Eq. (12))

it may also be related to a weaker polymer-filler coupling between the carbon black surface and the E-SBR-chains [119].

The observed effect of mixing on aggregate size distribution has a pronounced influence on the mechanical properties of the composites. This can be quantified by considering the solid fraction Φ_p of primary aggregates that represents a measure for the "structure" of carbon blacks. It is given by the ratio between the solid volume and the overall aggregate volume. Then, with Eq. (11) one finds the following scaling relation with respect to the average diameter *d* of the aggregates:

$$\Phi_p = \frac{V_p}{\frac{\pi}{6}d^3} \sim d^{d_f - 3} \qquad (14)$$

A fractal analysis according to Eq. (14) of carbon black aggregates (N339) in ready mixed E-SBR-composites is depicted in Fig. 23. The predicted scaling behavior of the aggregate solid fraction with diameter can be observed, though there is a larger scattering of the data as compared to those in Fig. 19. The slope of the solid regression line yields $d_f \approx 2.33$. Note that this differs significantly from the above estimate (Fig. 19a), which is mainly due to the different averaging procedures. An analysis of the solid fraction Φ_P, calculated according to Eq. (14), indicates that for a constant aggregate diameter d there is a distribution of Φ_P, as well as for a constant Φ_P there is a scatter of the average diameter. The limiting value of the solid fraction $\Phi_p=1$, corresponding to spherical particles, fits very well to the size of the primary particles of the carbon black N339 given by about 30 nm (dashed lines).

As argued above, aggregates are efficiently diminished in size and partly broken due to the higher shear forces with increasing carbon black loading. This is emphasized in Fig. 24 by the increasing value of the mean solid frac-

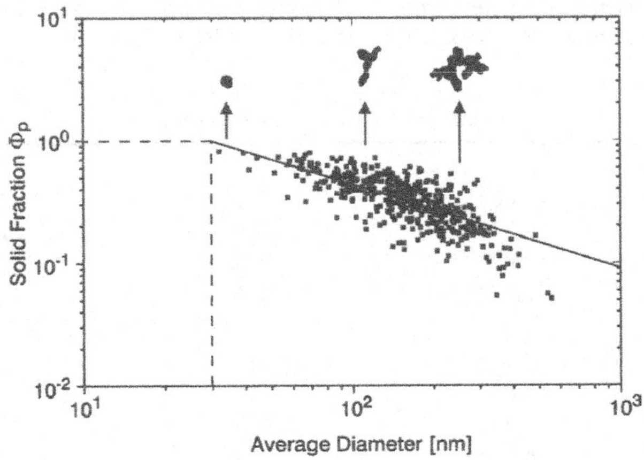

Fig. 23 Fractal analysis according to Eq. (14) of primary carbon black aggregates (N339) prepared from E-SBR-composites with 60 phr filler (in-rubber state). V_p is evaluated from Eq. (12)

tion $<\Phi_P>$ as the carbon black loading changes from 0.1 to 0.3 volume fraction. The different extent of aggregate breakdown in E-SBR- and S-SBR composites, summarized in Table 2, again becomes apparent. It is indicated by the two regression lines that show a slightly steeper increase in the case of S-SBR-composites (dashed line).

A technologically important advantage of a high primary aggregate "structure" (low solid fraction) is the improved dispersion behavior of the "high structure" blacks [26, 119]. In particular, the very fine blacks can hardly be dispersed by mechanical mixing due to the large number of attractive contacts between adjacent primary aggregates in a more or less close

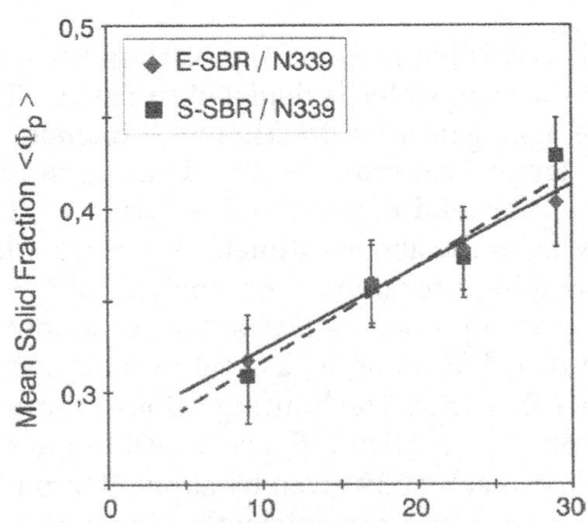

Fig. 24 Mean aggregate solid fraction vs filler volume fraction of N339 in E-SBR- and S-SBR-composites, respectively, as obtained from TEM-analysis (in-rubber state)

packing before mixing, e.g., in the pelletized state. With increasing "structure" (decreasing solid fraction), the number of attractive contacts per unit volume decreases and the force, necessary to separate the aggregates in a pellet, is reduced. In comparison, for the coarse blacks the number density of contacts between neighboring aggregates in a pellet is relatively small and therefore also "low structure" grades are quite easily dispersed.

Summarizing we conclude that the primary aggregate "structure" can be well analyzed by TEM-techniques. It is readily described by a fractal approach that refers to a cluster-cluster aggregation mechanism of primary particles during carbon black processing. During compounding of carbon blacks with highly viscous rubbers, rupture of primary aggregates takes place depending on the mixing severity. This leads to a reduced aggregate size and an increased solid fraction with increasing mixing time or filler loading, which has a significant influence on the mechanical properties of the composites. This will be considered in a micro-mechanical model of rubber reinforcement in Sect. 5.

4 Carbon Black Networking on Mesoscopic Length Scales

4.1 Microscopic and Scattering Analysis

A key factor for a deeper understanding of structure-property relationships of high performance elastomers is the morphological arrangement of filler particles on mesoscopic length scales due to secondary aggregation, also referred to as filler flocculation, cluster formation or networking. Even though the presence of such secondary network structures is often used in discussing filler specific mechanisms as the Payne effect, its morphology, as depicted in Fig. 25, is rarely studied on a quantitative level in comparison to that of the primary aggregates.

In particular, the morphology of secondary filler networks is of high relevance for micro-mechanical models of rubber reinforcement, involving the original relaxed network structure as well as its stress-induced breakdown during deformations. Such models have been developed by assuming, e.g., a percolation structure for the filler network [56] with a characteristic mass fractal dimension d_f=2.5 [6, 7]. Investigations of carbon black network structure in polyethylene (HDPE) by electric force microscopy confirm this value, approximately, delivering d_f=2.6 on length scales between 0.8 and 2 μm [113]. However, recent studies of carbon black (N330) network structure in rubber (EPDM) by scanning electron microscopy found significantly smaller values for the mass fractal dimension in the intermediate length scale range between about 200 nm and 2 μm, increasing from d_f=1.9 to d_f=2.4 with increasing filler concentration [120]. The variation of the fractal dimension was referred to a spatial inter-penetration of spanning cluster arms in three-dimensional space that does not allow for a proper estimation

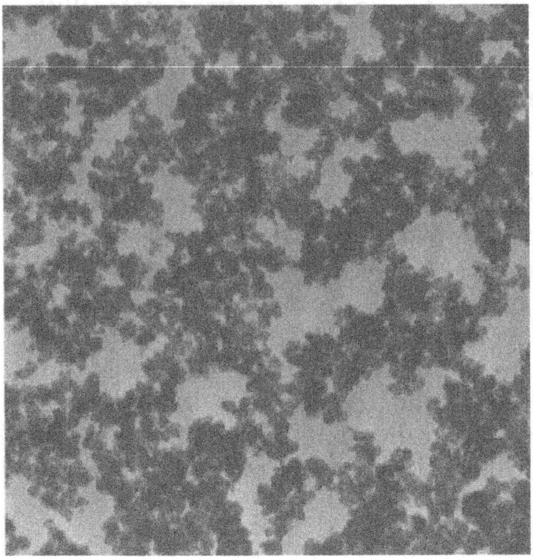

Fig. 25 TEM-micrograph of a carbon black network obtained from an ultra-thin cut of a filled rubber sample

of d_f. The value d_f=1.9 for low filler concentrations was conjectured to lie close to the characteristic value d_f=1.8 expected for cluster-cluster aggregation (CCA) [6, 7]. The CCA-morphology has been considered by Witten et al. in modeling the quasi-static stress-strain behavior of filler reinforced rubbers [21]. Furthermore, based on the assumption of a CCA-filler network, the scaling behavior of the small strain elastic modulus with filler concentration could be well described [22, 23, 48, 85], though there is no unique experimental evidence for this particulate morphology (see below).

So far, scattering techniques fail to provide reliable information on the connectivity of the filler network on larger length scales. This is again due to the spatial inter-penetration of the filler network in three-dimensional space, since labeling techniques, as considered, e.g., for polymer networks, cannot be applied. SAXS-studies on carbon black (N330) filled polyethylene have demonstrated that inter-penetration of primary aggregates in dense pellets causes the fractal domain to vanish due to a loss of correlation between primary particles [95]. A well defined scaling behavior of the scattering curves could only be obtained if a vanishing inter-penetration of neighboring aggregates was insured, i.e., spatially well separated aggregates, obtained from de-pelletized or fluffy black at sufficiently low filler concentrations, were realized. The same effect of inter-cluster correlation causes difficulties in obtaining information on the structure and connectivity of filler networks in elastomers, since inter-penetration cannot be avoided. This is also obvious in the TEM-micrograph shown in Fig. 25, where the three-dimensional, inter-penetrating connectivity of the carbon black network structure becomes apparent.

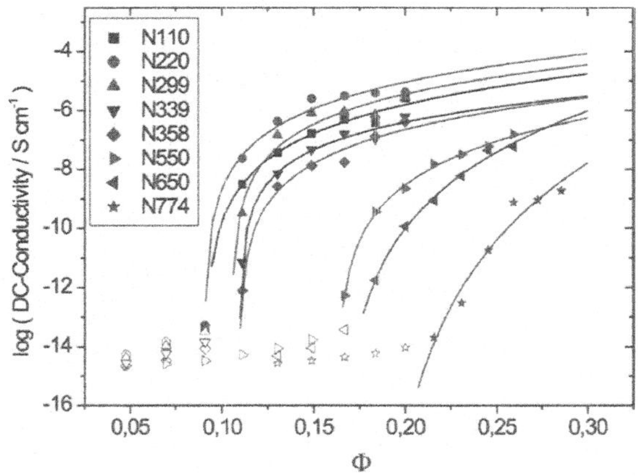

Fig. 26 D.c.-conductivity vs carbon black volume fraction of E-SBR samples filled with various furnace blacks. The *solid lines* are least square fits according to Eq. (15). Experimental data are taken from [124]

4.2
Investigations of Electrical Properties

4.2.1
Percolation Behavior of the d.c.-Conductivity

An important tool for analyzing the structure of carbon black networks in elastomers is provided by investigations of the electrical percolation behavior and the dielectric properties in a broad frequency range [22, 23, 121–139]. This is evident from the fact that, in an isolating polymer matrix, the charge carriers follow the conducting paths given by the connectivity of the filler network. In particular, the electrical percolation threshold decreases with increasing specific surface and/or "structure" of carbon black primary aggregates and decreasing compatibility between polymer and filler [121–125]. This emphasizes the role of mean aggregate distance or gap size between primary aggregates or clusters. It refers to a thermally activated hopping or tunneling of charge carriers across the gaps that governs the conductivity of carbon black filled polymers above the percolation threshold. The non-universal value of the percolation exponent of the d.c.-conductivity gives a further hint on the role of charge carrier tunneling in the conduction mechanism of filled polymer composites [126, 127].

The electrical percolation behavior for a series of carbon black filled rubbers is depicted in Fig. 26 and Fig. 27. The inserted solid lines are least square fits to the predicted critical behavior of percolation theory, where only the filled symbols are considered that are assumed to lie above the percolation threshold. According to percolation theory, the d.c.-conductivity σ_{dc} increases with the net concentration $\Phi-\Phi_c$ of carbon black according to a power law [6,128]:

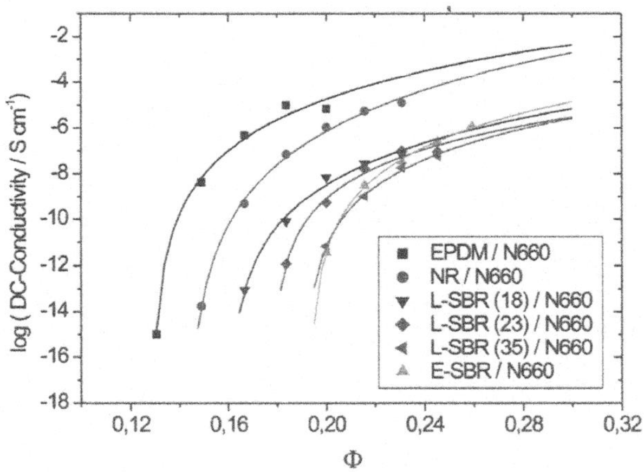

Fig. 27 D.c.-conductivity vs carbon black volume fraction of various rubber samples filled with the furnace black N660. The *solid lines* are least square fits according to Eq. (15). Experimental data are taken from [124]

$$\sigma_{dc} = \sigma_o \left(\frac{\Phi - \Phi_c}{1 - \Phi_c} \right)^{\mu} \quad for \ \Phi > \Phi_c \tag{15}$$

The critical concentration Φ_c denotes the percolation threshold, σ_o is the limiting d.c.-conductivity for Φ=1 and the exponent μ is called percolation exponent. As outlined in [23], percolation theory predicts a universal value $\mu \approx 2$ for all types of 3-D-lattices. However, the fitted lines in Fig. 26 and Fig. 27 yield a non-universal behavior of μ, i.e., significantly larger values that depend on the carbon black grade and on the type of rubber. This is apparent from the data shown in Table 3, where the fitting parameters Φ_c, μ, and σ_0 are listed together with the correlation coefficients r^2 for all sample series. Obviously, there is a clear trend that the critical concentration Φ_c as well as the percolation exponent μ increase with increasing grade number of carbon black and, apart from the very coarse blacks, the limiting conductivity σ_o decreases. Furthermore, these parameters are found to be highly affected by the microstructure of the rubber matrix.

This behavior can be understood if a superimposed kinetic aggregation process of primary carbon black aggregates in the rubber matrix is considered that alters the local structure of the percolation network. A corresponding model for the percolation behavior of carbon black filled rubbers that includes kinetic aggregation effects is developed in [22], where the filler concentrations Φ and Φ_c are replaced by effective concentrations. In a simplified approach, not considering dispersion effects, the effective filler concentration is given by:

$$\Phi_{eff} = \frac{\Phi}{\Phi_p} \left\{ 1 + \beta \left(\frac{\Phi - \Phi^+}{\Phi_p} \right)^B \right\} \tag{16}$$

Table 3 Fitting parameters of Eq. (15) for the data shown in Figs. 26 and 27

Sample series	N_2SA [m^2/g]	$\Phi_P(CDBP)$	Φ_c	μ	σ_0 [S/cm]	r^2
E-SBR/N110	138	0.36	0.091	3.7	3.9×10^{-3}	0.966
E-SBR/N220	116	0.36	0.090	3.4	1.3×10^{-2}	0.920
E-SBR/N299	106	0.35	0.105	3.2	4.2×10^{-3}	0.938
E-SBR/N339	95	0.35	0.110	2.7	2.0×10^{-4}	0.989
E-SBR/N358	85	0.34	0.109	3.4	4.8×10^{-4}	0.985
E-SBR/N550	41	0.41	0.162	4.0	8.0×10^{-4}	0.998
E-SBR/N650	38	0.40	0.162	7.1	0.36	0.994
E-SBR/N774	29	0.45	0.185	10.6	17.2	0.960
EPDM/N660	36	0.43	0.129	6.3	122	0.994
NR/N660	36	0.43	0.143	7.8	1078	0.998
L-SBR(18)/N660	36	0.43	0.159	6.2	0.43	0.992
L-SBR(23)/N660	36	0.43	0.178	4.6	0.026	0.986
L-SBR(35)/N660	36	0.43	0.191	5.2	0.093	0.997
E-SBR/N660	36	0.43	0.193	5.5	1.02	0.999

N_2SA: Nitrogen surface area
$\Phi_P(CDBP)$: Primary aggregate solid fraction from CDBP-absorption with data from [26]

In Eq. (16), Φ_P is the solid fraction of the primary aggregates, which takes into account that carbon blacks are "structured" (compare Sect. 3.2.). Φ^+ is a critical concentration (aggregation limit), where secondary aggregation starts and β and B are characteristic growth exponents of secondary aggregation. These three parameters are governed by the mean primary aggregate distance and their mobility in the rubber matrix, i.e., they depend on the specific surface and "structure" of carbon black grade and on the type of rubber. In particular, it is shown in [22] that the cluster growth function Eq. (16) allows for a quantitative description of the dependence of the Young modulus on filler concentration in the case of spherical, mono-disperse model fillers (micro-gels). In [48] it is demonstrated that the ability of secondary aggregation of filler particles in highly viscous polymers is reduced with increasing polymer-filler interaction. The same effect is described in the literature for the case of surface modified as well as graphitized carbon blacks [61, 129, 130].

The consideration of kinetic aggregation of "structured" filler particles by an effective filler concentration Φ_{eff}, as given by Eq. (16), instead of Φ and Φ_c in the percolation model Eq. (15), allows for a consistent interpretation of the observed variation of fitting parameters listed in Table 3. A first effect on Φ_c results from the "structure", as described by Φ_P, that generally tends to larger values for the coarse blacks with large grade number. Since TEM-data for Φ_P, shown, e.g., in Fig. 23 and Fig. 24, are not available for all sample series, the values $\Phi_P(CDBP)=(1+\rho_{CB}\ CDBP/100)^{-1}$ are inserted in Table 3. They are given by the amount of di-butylphthalate absorption per 100 g of carbon black after mechanical treatment (compressed DBP number ($CDBP$) according to ASTM D3493-90) and the mass density ρ_{CB}=1.8 g/cm^3 of carbon black. Obviously, a single factor, e.g., the numerically estimated value of the percolation threshold in a simple cubic lattice, $\Phi_{c,eff}\approx0.31$ [6], gives no correlation between $\Phi_P(CDBP)$ and the experimental values of Φ_c, indicating that the external bracket term of Eq. (16) differs from one. It means that the power law term, considering kinetic aggregation, must be significant. Accordingly, the percolation threshold Φ_c increases with increasing grade number mainly due to the restriction of mobility with decreasing specific surface of the primary aggregates.

The impact of rubber type on Φ_c (Fig. 27) can also be related to the variation of particle mobility, since a strong polymer-filler coupling in a more compatible polymer matrix reduces the particle mobility. Accordingly, beside the effect of "structure" considered by $\Phi_P(CDBP)$, the enhanced polymer-filler interaction in the series from EPDM via NR to the SBRs [124] as well as the decreasing specific surface area (N_2SA) with rising carbon black grade number lead to an increase of the percolation threshold Φ_c, since both effects hinder the kinetic aggregation process. A reduced aggregation corresponds to higher values of the growth exponent B in Eq. (16), which can be related to the increase of the percolation exponent μ with increasing grade number, observed in Table 3. This is realized on a qualitative level, when the percolation Eq. (15) is rewritten with effective volume fractions as given by Eq. (16). Then, if for sufficient large filler concentrations the one in the first bracket term of Eq. (16) is neglected and if furthermore Φ^+ is approximated by Φ_c, one obtains an approximative scaling equation for the conductivity with respect to the net concentration $\Phi-\Phi_c$ with an effective exponent $\mu(B+1)$ that can be significantly larger than μ. This explains the non-universal value of the conductivity exponent that increases with the cluster growth exponent B.

Consequently, the consideration of a superimposed kinetic aggregation in the framework of percolation theory allows for a qualitative understanding of the variation of Φ_c and μ for the different sample series. We finally note that for a proper quantitative description of the percolation behavior of carbon black composites it is also necessary to consider the effect of primary aggregate breakdown more closely, i.e., the dependence of Φ_P on filler concentration and mixing conditions entering Eq. (16) (compare Sect. 3.2.2). This is beyond the scope of the present chapter. A quantitative description of percolation including a full characterization of in-rubber morphology of primary carbon black aggregates will be a task of future work.

4.2.2
Dielectric Analysis of Carbon Black Networks

The dielectric properties of carbon black filled rubbers are closely related to the morphological structure of filler networks, providing an important analysis tool on mesoscopic length scales. On the one hand, the tunneling or hopping of charge carriers over conductive gaps provide information on the specific morphology of filler-filler bonds on nanoscopic length scales [121, 122, 131–134]. On the other hand, the scaling behavior of the conductivity

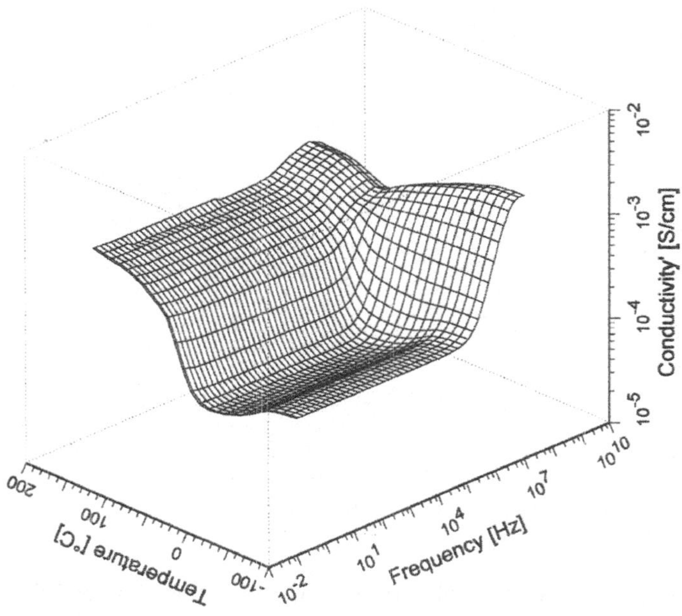

Fig. 28 A.c.-conductivity σ′ vs frequency and temperature of a NR-sample filled with 80 phr carbon black (N339)

in the high frequency regime is indicative for anomalous diffusion of charge carriers on fractal carbon black clusters on mesoscopic length scales [23, 134–139]. In the following, both of these effects will be analyzed by referring to exemplary results.

Fig. 28 shows the real part of the a.c.-conductivity σ' of a highly filled NR/N339 sample in the frequency- and temperature range 10^{-1} to 10^{7} Hz and −100 to 150 °C, respectively. At low frequencies, below around 10^{5} Hz, the conductivity is almost independent of frequency but shows a pronounced temperature dependence. At high frequencies, a relaxation transition is observed leading to a step-like increase of the conductivity. In the low frequency regime, the conductivity first decreases slightly with increasing temperature, when the glass transition temperature, $T_g \approx$-60 °C, is exceeded. This can be related to an enhanced thermal expansion of the rubber above T_g. It increases the size of conducting gaps between adjacent, highly conducting carbon black aggregates somewhat, implying that the gap size is the dominating factor for the conductivity. Since no thermal activation of the conductivity is observed at low temperatures, the conduction mechanism can be concluded to be due to tunneling across gaps. Consequently, the gap size is of the order of 1 nm, i.e., the typical tunnel distance of quantum particles [121, 122]. It can be related to the presence of a thin polymer film in the contact area between flocculated primary carbon black aggregates.

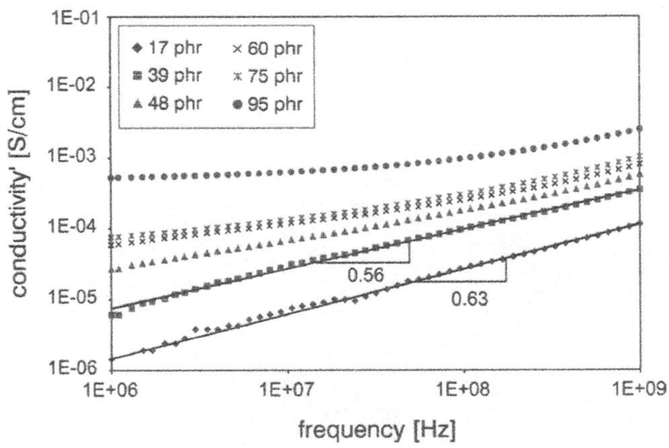

Fig. 29 A.c.-conductivity σ' vs frequency at 20 °C of NBR-samples filled with various amounts of N220, as indicated. The *solid lines* correspond to the scaling behavior Eq. (17), obtained in the high frequency range up to 1 GHz

With rising temperature, above around 0 °C, the conductivity shown in Fig. 28 increases significantly. A closer analysis indicates a characteristic Arrhenius behavior in the temperature regime between 20 and 60 °C. Recently, such behavior has also been observed for other carbon black filled elastomers [133]. Obviously, the conduction mechanism in this temperature regime changes, implying a thermally activated hopping of charge carriers across the gaps between adjacent carbon black aggregates [121, 122]. For temperatures larger than about 60 °C the conductivity levels out. This can be related to the impact of an increasing gap size on the conductivity, i.e., the strong exponential decrease of the conductivity with increasing gap size due to a more pronounced thermal expansion. Investigations of highly filled composites with various rubbers show that a qualitative similar behavior of the temperature dependence of the conductivity as in Fig. 28 is realized, but the temperature where the conductivity levels out seems to be roughly correlated with the glass transition temperature of the polymer. In particular, for BR-composites with a relatively low glass transition temperature, the conductivity levels out around room temperature and shows a pronounced drop by orders of magnitude at higher temperatures. For these systems a conductor-isolator transition is observed at about 130 °C, which must be related to a strong increase of the gaps beyond a distance where hopping or tunneling of charge carriers can take place.

For a deeper understanding of structure-property relationships it is useful to consider the effect of carbon black grade and concentration as well as polymer type on the dielectric properties more closely. In Fig. 29 the real part of the a.c.-conductivity σ' at 20 °C of a series of rubber composites, consisting of the more polar statistical co-polymer NBR and the fine black N220, is depicted for various filler concentrations in the high frequency regime up to 1 GHz. For the lower carbon black concentrations, a power law behavior with exponent around 0.6 is observed, while the highly filled com-

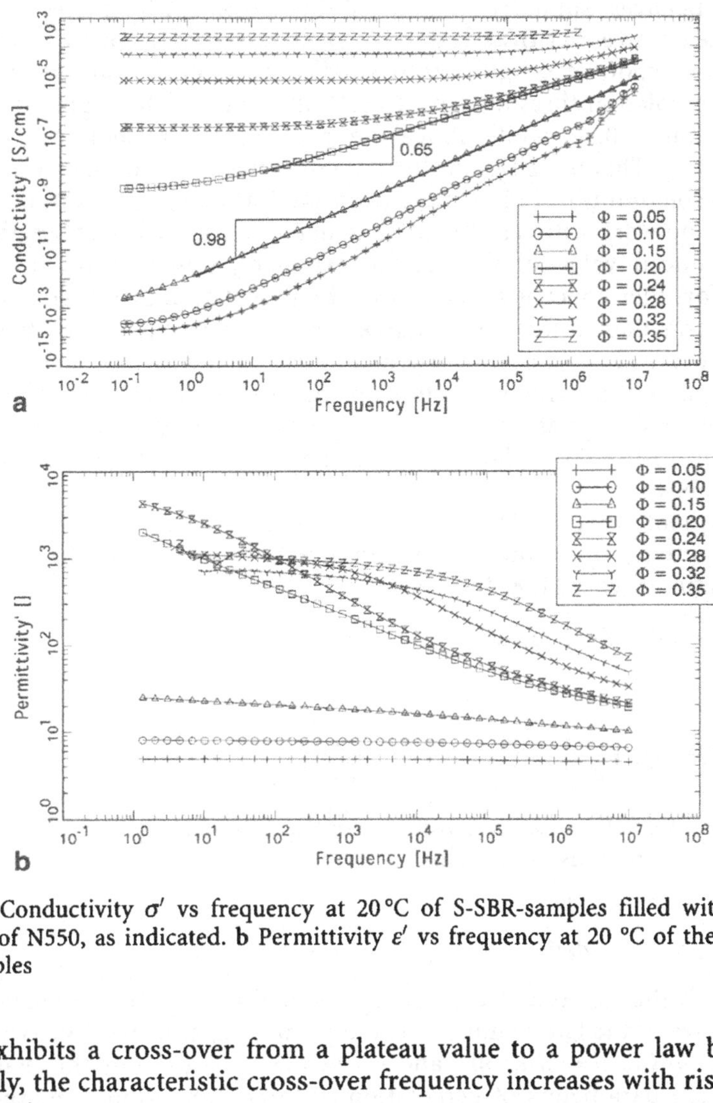

Fig. 30 **a** Conductivity σ' vs frequency at 20 °C of S-SBR-samples filled with various amounts of N550, as indicated. **b** Permittivity ε' vs frequency at 20 °C of the same S-SBR-samples

posite exhibits a cross-over from a plateau value to a power law behavior. Obviously, the characteristic cross-over frequency increases with rising filler concentration.

This behavior becomes more transparent in Fig. 30a,b, where the a.c.-conductivity σ' and relative dielectric constant (permittivity ε'), respectively, for a series of less polar S-SBR-samples filled with various amounts of the coarse black N550 are show at 20 °C in a broader frequency range up to 10^7 Hz. For filler concentrations below the percolation threshold ($\Phi \leq 0.15$), the conductivity behaves essentially as that of an isolator and increases almost linearly with frequency. Above the percolation threshold ($\Phi \geq 0.2$), it shows a characteristic conductivity plateau in the small frequency regime. Since at low frequencies the value of the conductivity σ' agrees fairly well with the d.c.-conductivity, the plateau value exhibits the characteristic percolation behavior considered above. In the high frequency regime the conductivity depicted in

Fig. 30a behaves similarly to that of the NBR/N220-samples shown in Fig. 29, i.e., above a critical frequency ω_ξ it increases according to a power law with an exponent n significantly smaller than one. In particular, just below the percolation threshold for Φ=0.15 the slope of the regression line in Fig. 30a equals 0.98, while above the percolation threshold for Φ=0.2 it yields n=0.65. This transition of the scaling behavior of the a.c.-conductivity at the percolation threshold results from the formation of a conducting carbon black network with a self-similar structure on mesoscopic length scales.

The reduced value of the scaling exponent, observed in Fig. 29 and Fig. 30a for filler concentrations above the percolation threshold, can be related to anomalous diffusion of charge carriers on fractal carbon black clusters. It appears above a characteristic frequency ω_ξ, when the charge carriers move on parts of the fractal clusters during one period of time. Accordingly, the characteristic frequency ω_ξ for the cross-over of the conductivity from the plateau to the power law regime scales with the correlation length ξ of the filler network.

For a quantitative analysis of the scaling and cross-over behavior of the a.c.-conductivity above the percolation threshold we refer to the predictions of percolation theory [128, 136, 137]:

$$\sigma'(\omega) \sim \omega^n \quad \text{for} \quad \omega > \omega_\xi \tag{17}$$

and

$$\omega_\xi(\Phi) \cong \omega_{\xi,o}\left(\frac{\xi}{d_o}\right)^{-d_w} \cong \omega_{\xi,o}\left(\frac{\Phi - \Phi_c}{1 - \Phi_c}\right)^m \tag{18}$$

For the evaluation of the front factor the Einstein equation for the conductivity σ_o can be used. It yields

$$\omega_{\xi,o} = \tau_{\xi,o}^{-1} = \frac{6\sigma_o k_B T}{e^2 n_o d_o^2} \tag{19}$$

Here, e is the electron electric charge, n_o is the charge carrier density, and d_o is the size of the lattice units, i.e., the primary aggregates. The two critical exponents $n=(d_w-d_f+1)/d_w\approx0.6$ and $m=\nu d_w\approx3.3$ are given by the characteristic structure parameters of percolation in three dimensions, i.e., the fractal dimension $d_f\approx2.5$, the anomalous diffusion exponent $d_w\approx3.8$, and the correlation length exponent $\nu\approx0.87$ [128, 137]. The experimental value n=0.65 found in Fig. 30a is in fair agreement with this prediction, indicating that for sufficient high carbon black concentrations a percolation structure of the filler network is realized in the systems. However, the experimental value n=0.65 lies also not far away from the predicted value $n\approx0.74$ for a kinetically aggregated CCA-network structure, obtained with a fractal dimension $d_f\approx1.8$ and an anomalous diffusion exponent $d_w\approx3.1$. A more detailed discussion of the scaling behavior of the conductivity in the framework of the two models, percolation and kinetic aggregation, is presented in [23].

The permittivity shown in Fig. 30b characterizes the polarization of the samples in an alternating field. At low carbon black concentrations, the permittivity is relatively small ($\varepsilon' \cong 10$) and almost independent of frequency. At higher carbon black concentrations, above the percolation threshold ($\Phi \geq 0.2$), relatively high values of the permittivity ($\varepsilon' \cong 10^3$) are found in the low frequency regime, but with increasing frequency a relaxation transition takes place and the permittivity ε' falls off drastically. The location of the relaxation transition on the frequency scale, ω_R, is shifted to higher frequencies with rising carbon black concentration. Note that the corresponding relaxation current leads to the step-like increase of the conductivity in the high frequency regime, observed for the system NR/N339 in Fig. 28. For the present sample series S-SBR/N550, shown in Fig. 30a,b, this relaxation current is not significant on the chosen logarithmic scale. Obviously, it is hidden by the relatively high conduction plateau. In this sense, conductivity and permittivity are decoupled, allowing for a proper analysis of the anomalous diffusion- and polarization transitions, separately.

An explanation of the observed relaxation transition of the permittivity in carbon black filled composites above the percolation threshold is again provided by percolation theory. Two different polarization mechanisms can be considered: (i) polarization of the filler clusters that are assumed to be located in a non polar medium, and (ii) polarization of the polymer matrix between conducting filler clusters. Both concepts predict a critical behavior of the characteristic frequency ω_R similar to Eq. (18). In case (i) it holds that $\omega_R \cong \omega_\xi$, since both transitions are related to the diffusion behavior of the charge carriers on fractal clusters and are controlled by the correlation length ξ of the clusters. Hence, ω_R corresponds to the anomalous diffusion transition, i.e., the cross-over frequency of the conductivity as observed in Fig. 30a. In case (ii), also referred to as random resistor-capacitor model, the polarization transition is affected by the polarization behavior of the polymer matrix and it holds that [128, 136, 137]

$$\omega_R(\Phi) \cong \frac{1}{R_o C_o} \left(\frac{\Phi - \Phi_c}{1 - \Phi_c} \right)^q \tag{20}$$

Here, R_o is the resistance of the occupied lattice units and C_o is the capacitance of the non-occupied lattice units. In a first attempt, not considering anomalous diffusion effects, the exponent has been evaluated as $q=\mu+s\approx 2.7$, where $s\approx 0.7$ and $\mu\approx 2.0$ are the conductivity exponents (in three dimensions) below and above the percolation threshold, respectively. However, by including anomalous diffusion effects one obtains $q=m=\nu d_w\approx 3.3$ [136, 137].

For a quantitative analysis of the percolation behavior of ω_R, considered in Eq. (20), the permittivity data shown in Fig. 30b have been fitted to an empirical Cole-Cole function [131, 132]. The fits are quite good, yielding relatively small broadness parameters between 0.36 and 0.5. The obtained relaxation times $\tau_R=\omega_R^{-1}$ are depicted in Fig. 31 in dependence of filler concentration Φ. Furthermore, the cross-over times $\tau_\xi=\omega_\xi^{-1}$ and the limiting low frequency plateau values of the conductivity σ' ($\omega\to 0$), obtained from the data in Fig. 30a, are represented in Fig. 31. They have been evaluated by a simple shifting procedure for constructing a conductivity master curve, as

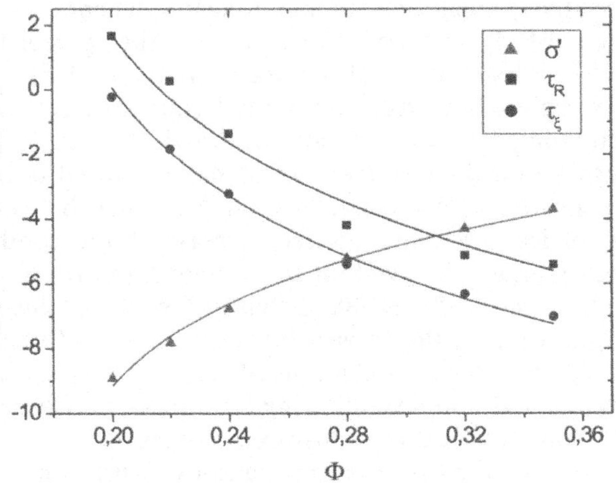

Fig. 31 Characteristic parameters of the S-SBR/N550 composites, σ' ($\omega\to 0$), τ_ξ, and τ_R, for various carbon black concentrations Φ obtained from Fig. 30a, b. The *solid lines* are adapted to Eqs. (15), (18), and (20), respectively

applied, e.g., in [135]. The solid lines depicted in Fig. 31 are fitted curves for the characteristic composite parameters σ' ($\omega\to 0$)$=\sigma_{dc}$, τ_ξ, and τ_R according to Eqs. (15), (18) and (20), respectively. Thereby, a common exponent $q=m=10.1$ and a single value of the percolation threshold $\Phi_c=0.165$ have been assumed, providing best fits to the experimental data with correlation coefficients r^2 between 0.978 and 0.996. The conductivity exponent corresponding to Eq. (15) is found as $\mu=7.4$ and the limiting conductivity for $\Phi=1$ yields $\sigma_o=12.4$ S/cm. For the two front factors of Eqs. (18) and (20) one obtains $\tau_{\xi,o}=1.4\times10^{-14}$ s and $\tau_{R,o}=6.3\times10^{-13}$ s.

If the estimated fitting parameters are compared to the predicted values of percolation theory, one finds that all three exponents are much larger than expected. The value of the conductivity exponent $\mu=7.4$ is in line with the data obtained in Sect. 3.3.2, confirming the non-universal percolation behavior of the conductivity of carbon black filled rubber composites. However, the values of the critical exponents $q=m=10.1$ also seem to be influenced by the same mechanism, i.e., the superimposed kinetic aggregation process considered above (Eq. 16). This is not surprising, since both characteristic time scales of the system depend on the diffusion of the charge carriers characterized by the conductivity.

The front factors obtained from the fittings in Fig. 31 can be used for a dielectric characterization of the network units, e.g., for an evaluation of the charge carrier density n_o of the primary carbon black aggregates as given by Eq. (19). Therefore, the electric charge $e^2=2.3\times10^{-28}$ Jm and the conductivity $\sigma_o=1.1\times10^{13}$ s^{-1} have to be inserted with respect to Gaussian units. The mean diameter of primary aggregates in S-SBR composites (50 phr N550) has been obtained from TEM analysis as $d_o=157$ nm. With $T=293$ K and $k_B=1.38\times10^{-23}$ J/K this yields for the reduced charge carrier density n_o

d_o^3=2.6, i.e., the number of charge carriers per primary aggregate is of order one. This relative small number of charge carriers can be related to the tunneling or hopping conductivity mechanism between conducting nanoscopic islands (primary aggregates) with a high Coulomb exclusion energy. The exclusion energy for a single electron can be roughly estimated from the limiting relaxation time $\tau_{R,o}=R_oC_o$, if the resistance R_o of the lattice units is expressed by the conductivity σ_o via $R_o\cong(\sigma_o d_o)^{-1}=5.1\times10^3\ \Omega$. This yields for the capacitance of the lattice units $C_o=1.2\times10^{-16}$ F, or in Gaussian units $C_o=1.1\times10^{-6}$ m. Hence, the exclusion energy of a single electron results as $E_C=e^2/2C_o=10^{-22}$ J. This value is more than one order of magnitude smaller than the thermal energy $k_BT=4\times10^{-21}$ J, which has to be exceeded for insuring the dominance of the Coulomb exclusion energy. Accordingly, from this estimates it appears that the charge carrier density is not limited by the exclusion energy but governed by thermal fluctuations. However, note that this calculation is not very precise, since the local geometrical structure of the lattice units is not considered properly. A more fundamental study of charge transport mechanisms in carbon black filled rubbers should take into account that the lattice units are fractal objects (primary carbon black aggregates), which are separated by small gaps not larger than a few nanometers.

4.3 Flocculation Dynamics and the Nature of Filler-Filler Bonds

For a deeper understanding of filler networking in elastomers it is useful to monitor structural relaxation phenomena during heat treatment (annealing) of the uncross-linked composites. This can be achieved by investigations of the time development of the small strain storage modulus G'_0 that provides information about the flocculation dynamics [140–143]. Figure 32a shows the time development of G'_0 of S-SBR melts of variable molar mass filled with 50 phr carbon black (N234), when a step-like increase of the temperature from room temperature to 160 °C is applied. Figure 32b shows a strain sweep of the same systems after 60 min annealing time. Dependent on the molar mass M_w, as indicated, a pronounced increase of G'_0 is observed in the first minutes that levels out more or less to a plateau value at larger annealing times. In agreement with recent studies of Wang et al. [143], the largest plateau value is observed for the lowest molar mass, indicating that the increase of the modulus results from flocculation of primary aggregates to form secondary aggregates (clusters) and finally a filler network. It appears that a weakly bonded superstructure develops in the systems during heat treatment that stiffens the polymer matrix. With increasing dynamic strain amplitude, as depicted in Fig. 32b, a stress-induced breakdown of the filler clusters takes place and the storage modulus decreases by about one order of magnitude (Payne effect). With respect to the variable molar mass of the systems, Fig. 32b shows a cross-over of the moduli with increasing strain, indicating that a larger molar mass stabilizes the filler-filler bonds more effectively. This can be related to the overlapping action of tightly bound polymer chains in the contact area between adjacent filler particles.

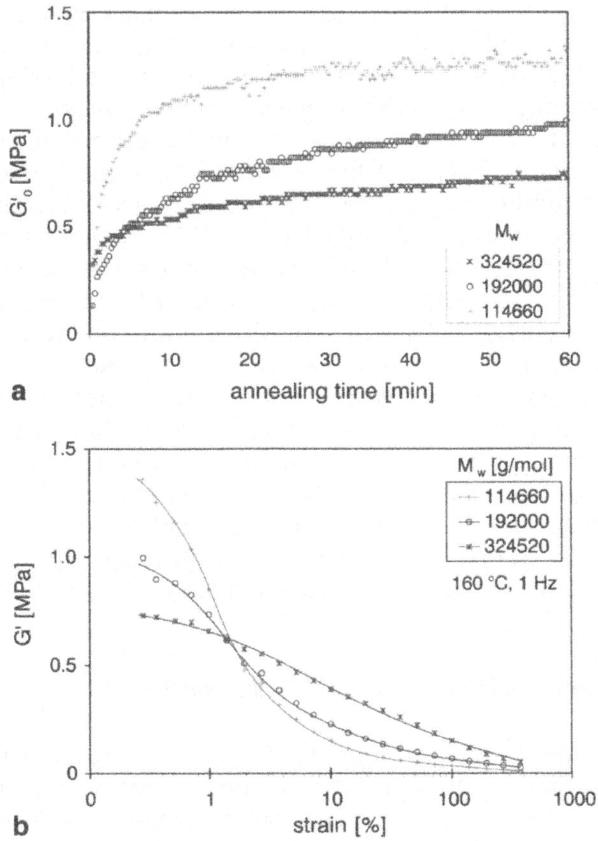

Fig. 32 **a** Time development of the small strain storage modulus of uncross-linked S-SBR composites with 50 phr N234 during heat treatment at 160 °C for various molar masses, as indicated (0.28% strain, 1 Hz). **b** Strain dependency of the storage modulus of the samples depicted in a after heat treatment for 60 min at 160 °C

Inspired by the above flocculation analysis and the dielectric investigations described before, a model concerning the structure-property relationships of filler-filler bonds in a bulk rubber matrix can be developed as illustrated in Fig. 33. According to this model, the stiffness of filler-filler bonds is governed by the remaining gap size between contacting particles. This, in turn, depends on the ability to squeeze out the bound polymer chains from the contact area under the attractive action of the van der Waals force between the filler particles. This process leads to a stiffening of filler-filler bonds. It is favored by several factors, e.g., a high ambient temperature, low molar mass, small particle size, weak polymer-filler- and strong filler-filler interaction.

The mechanical connectivity between the filler particles is provided by a flexible, nanoscopic bridge of glassy-like polymer, resulting from the immobilization of the rubber chains in the confining geometry close to the gap.

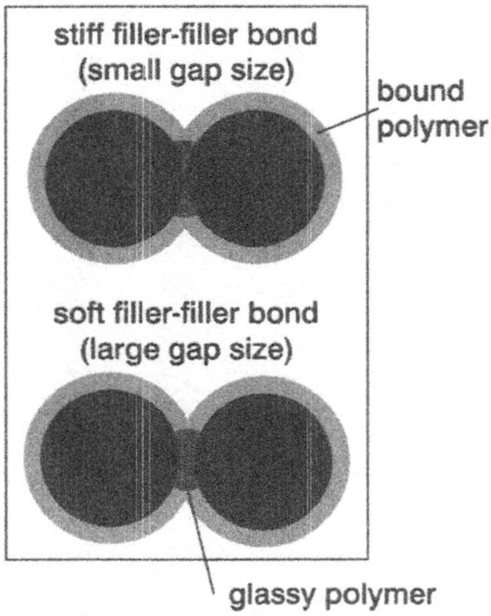

Fig. 33 Schematic view considering the structure of filler-filler bonds in a bulk rubber matrix. The impact of gap size on the stiffness of filler-filler bonds becomes apparent

Since the stiffness of the bonds transfers to the stiffness of the whole filler network, the small strain elastic modulus of highly filled composites is expected to reflect the specific properties of the filler-filler bonds. In particular, the small strain modulus increases with decreasing gap size during heat treatment as observed in Fig. 32a. Furthermore, it exhibits the same temperature dependence as that of the bonds, i.e., the characteristic Arrhenius behavior typical for glassy polymers. Note however that the stiffness of the filler network is also strongly affected by its global structure on mesoscopic length scales. This will be considered in more detail in the next section.

In the case of carbon black filled diene-rubber composites the polymer-filler interaction is generally quite strong due to the high affinity between the π-electrons at the carbon surface and those in the double bonds of the chains. According to the site energy distribution function estimated in Sect. 3.1.2, the typical interaction energy between carbon black and ethylene, representing a single double bond, lies between 10 and 35 kJ/mol and depends on the grade number. A more practical procedure for characterizing the polymer-filler interaction in elastomer composites is the estimation of bound rubber, i.e., the amount of polymer tightly bound to the filler surface after mixing [144]. It is well known that this amount increases with the molar mass of the polymer and the specific surface area of the filler particles, but is also affected significantly by the surface activity, given, e.g., by the site energy distribution function of the filler obtained with polymer analogous gases [144–147]. A further effect comes in from the preparation conditions of the composites, e.g., mixing time [148], since the formation of bound rubber is a slow dynamical process that requires time.

Figure 34 shows the result of bound rubber estimates at room temperature for variously mixed S-SBR composites filled with 50 phr of four different carbon blacks. The furnace blacks were mixed for 3 and 6 min in an in-

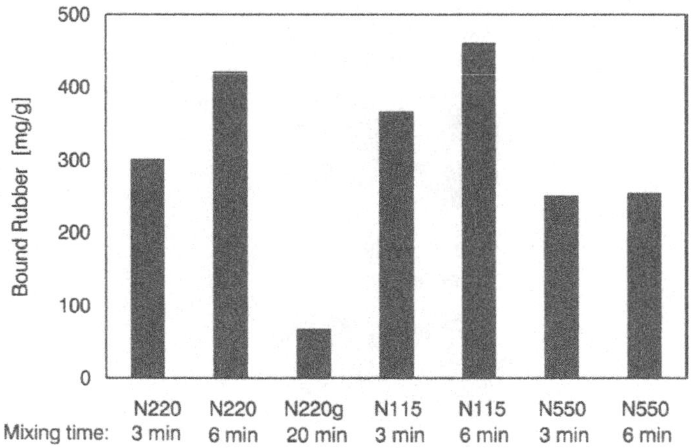

Fig. 34 Results of bound rubber estimates at room temperature for various carbon black filled S-SBR composites, as indicated

ternal mixer, respectively, while the graphitized black N220g was mixed for 20 min on a roller mill. The dispersion rating was found to be larger than 90% for all systems. It becomes obvious that the amount of bound rubber increases with increasing mixing time from 3 to 6 min. This results from an exchange of shorter chains, already bound to the filler surface, by longer chains during the ongoing mixing process. The longer chains are bound more tightly to the surface due to a higher number of adhesive contact spots per chain [144–146]. Beside the effect of mixing time the estimated bound rubber reflects the difference in specific surface of the furnace blacks that increases from N550 (41 m^2/g) via N220 (116 m^2/g) to N115 (143 m^2/g). A deviation from this correlation is observed for the graphitized black N220g that shows a significantly smaller amount of bound rubber as compared to its relatively large specific surface (88 m^2/g). This effect can be attributed to the lower surface activity of the graphitized black that results from the marked reduction of highly energetic sites during graphitization (compare Fig. 15 and Table 1). From this observation it appears that the highly energetic sites are the dominating factor for strong polymer-filler couplings.

The different interaction strength between polymer and filler leads to characteristic implications for the structural relaxation behavior (flocculation) of filler particles in highly entangled polymer melts. Figure 35 shows the strain dependency of the storage modulus G' for the two composites S-SBR/N220g and S-SBR/N220 (mixing time 6 min) before and after heat treatment (annealing) of the samples for 20 min at 160 °C. Obviously, a significant increase of the small strain modulus of both systems results during heat treatment, referring to the flocculation of filler particles. With increasing strain the moduli decrease significantly due to a stress-induced successive breakdown of filler clusters. A comparison of the two systems filled with the furnace black N220 and the graphitized black N220g shows a stronger increase of the moduli for the system with the graphitized black, indicating a

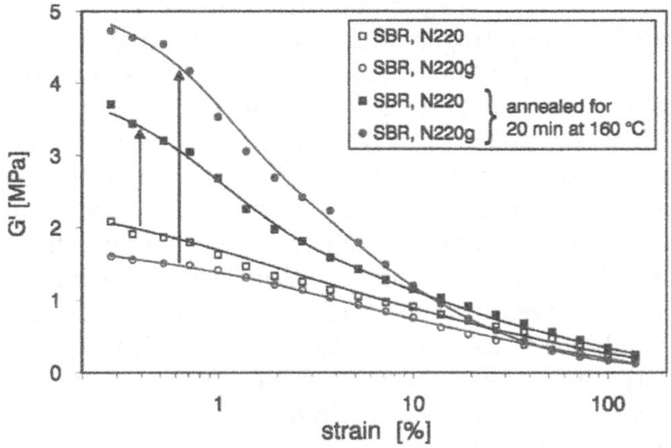

Fig. 35 Strain dependency of the storage modulus (G' at 60 °C) of uncross-linked S-SBR composites filled with 50 phr N220 and N220g, respectively, before and after heat treatment for 20 min at 160 °C

more pronounced flocculation during heat treatment. According to the model of filler-filler bonds considered in Fig. 33, the lower amount of tightly bound polymer around the graphitized black N220g supports the flocculation. It implies smaller contact gaps, because the polymer between the contacting particles can be squeezed out more readily by the attractive van der Waals forces. This leads to stiffer filler-filler bonds and hence to a higher modulus of the filler network. From principal reasons it is also possible that the global structure of the filler network of the two systems prepared with the furnace and graphitized black after heat treatment is generally different, e.g., the arrangement of the graphitized carbon particles is more compact. However, we will argue below that the formation of the filler network is governed by a diffusion controlled cluster-cluster aggregation (CCA)-process that leads to a universal CCA-network structure. Under this condition the difference in stiffness of the two systems has to be related to a change of the local stiffness of the filler-filler bonds.

The impact of bound rubber on the spacing of filler-filler contact gaps during annealing of carbon black composites at elevated temperature is affected by the amount of bound rubber estimated at this particular temperature. However, the values listed in Fig. 34, obtained at room temperature, may not be representative, because the amount of bound rubber depends on the extraction temperature. In general, the amount of bound rubber is also affected by the solvent used for extraction and decreases with increasing extraction temperature [150–153]. In earlier studies a linear decrease in bound rubber with temperature was observed, extrapolating to zero at 375 °C [150]. Other authors found a much more dramatic decrease with minimum bound rubber at about 100 °C [151]. Subsequent studies of Chapman et al. [152], comparing extraction in vacuum, nitrogen, and air, respectively, found a moderate decrease of bound rubber of about 30% at 100 °C in vacuum as compared to room temperature and a much more pronounced decrease in air and also in nitrogen atmosphere. They concluded that oxidative chain scission reduces the bound rubber content at elevated temperature if determined in air and also in nitrogen. Similar results were obtained by Kida et al. [153]. They concluded from NMR-relaxation measurements that only the content of the outer, most mobile rubber phase is reduced while the highly immobilized phase close to the carbon black surface remains almost constant up to 113 °C. From these results we can expect that for an annealing temperature of about 160 °C, which is also applied during vulcanization, bound rubber is still present at the carbon black surface and hinders the aggregation of neighboring particles. Note that this is also expected from simple energetic arguments, since the above estimated adsorption energy of ethylene on N220, $Q \approx 10-35$ kJ/mol, is almost one order of magnitude larger that the thermal energy RT=3.6 kJ/mol at 160 °C.

The effect of bound rubber on the stiffness of filler-filler bonds is essential for the small strain modulus G'_0. It allows for a qualitative explanation of various well known phenomena described, e.g., in the pioneering papers of Payne [28–35]. In particular, the observed decrease of G'_0 with increasing mixing time [33, 148] can be related to a softening of the bonds due to a successive increase of bound rubber. Furthermore, the variation of G'_0 with polymer type [30, 154] can be related to the impact of bound rubber on the spacing of filler-filler bonds, implying that a strong polymer-filler coupling or high molar mass lowers the value of G'_0. Note that this interpretation can also explain the more or less significant correlation between the small strain modulus and the electrical conductivity level obtained, e.g., in [154]. This kind of correlation is also found, if composites with graphitized and non-graphitized carbon blacks of the same grade number are compared. The higher stiffness of composites with graphitized blacks is typically accompanied by a higher conductivity level [121]. Both effects refer to a smaller gap size between contacting primary aggregates in the case of graphitized blacks.

Beside the stiffness of filler-filler bonds, the amount of bound rubber impacts also the strength of the bonds between interacting filler particles. Due to the overlapping action of the tightly bound polymer chains close to the contact area of neighboring filler particles, the mechanical stability of filler-filler bonds increases with increasing amount of bound rubber. This explains the cross-over behavior of the moduli (after heat treatment) as observed in Fig. 32b and Fig. 35: The systems with the lower amount of bound rubber, i.e., the composites with graphitized blacks and the low molar mass of the polymer, respectively, show the largest values of the small strain modulus G'_0, but the drop of the moduli with increasing strain appears at significantly smaller strain amplitudes. The stabilizing bound rubber layer can be considered to act like a spanning net around the filler clusters, implying a high flexibility of the filler network. We will see that this kind of reinforcing action is responsible for the pronounced hysteresis and high strength of reinforced rubbers in the dynamic strain regime up to around 100%. Further significant effects of the bound rubber layer result in the quasi-static high strain regime. This will be considered in the next section.

In conclusion, the investigations of electrical properties of carbon black filled rubbers, presented in Sect. 4.2, indicate that a percolation structure for the filler network is realized on mesoscopic length scales, if a critical filler concentration is exceeded. This structure seems to be modified by a superimposed kinetic aggregation process that refers to the non-universal value of the percolation exponent and the impact of specific surface on the percolation threshold. Further experimental evidence for a kinetic aggregation mechanism of colloidal particles dispersed in a rubber matrix is given by the mechanical response of the uncross-linked composites during heat treatment (annealing), demonstrating that a relative movement (flocculation) of the particles takes place that depends on particle size, molar mass of the polymer as well as polymer-filler and filler-filler interaction. It has been argued that the mechanical stiffness of filler-filler bonds is governed by the remaining gap size between adjacent filler particles that develops during annealing (and cross-linking) of filled rubbers. Based on the model of filler-filler bonds depicted in Fig. 33, a qualitative explanation of the observed flocculation effects is possible by referring to the amount of bound rubber and its impact on the stiffness and strength of filler-filler bonds.

5
Rubber Reinforcement by Fractal Filler Networks

In the last decades, remarkable progress has been obtained in understanding filler networking and its implications on the mechanical response of dynamically excited and highly strained rubber composites. This is summarized in a recent review of Heinrich and Klüppel [57]. According to this paper, two fundamental micro-mechanical concepts of non-linear viscoelasticity, which are based on fractal approaches of filler networking, have been developed: the (L-N-B)-model [56] and the cluster-cluster-aggregation (CCA)-model [21–23, 85]. They consider the arrangement of filler particles in clusters with well defined fractal structure and the elasticity or fracture of such clusters under external strain. The two models refer to different geometrical arrangements of filler particles in particulate fractal network structures, described by percolation theory or kinetic cluster-cluster aggregation, respectively.

We will focus here on the CCA-model of filler networking, since the investigations of flocculation and electrical properties considered in the last section provide strong evidence for a kinetic aggregation mechanism of filler particles in elastomers. A further strong indication for the CCA-mechanism is given by the predicted scaling behavior of the small strain modulus considered below. We will see that this scaling behavior is well fulfilled for different elastomer systems with various fillers. On the contrary, the corresponding scaling prediction of percolation theory clearly fails if applied for filler reinforced rubbers, since experimental data result in much smaller exponents than the predicted one [57].

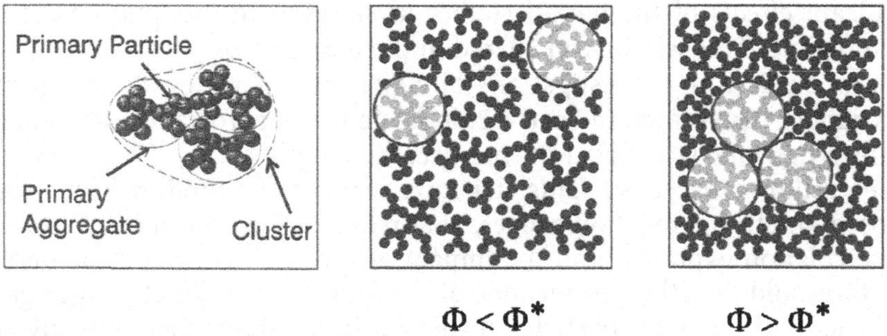

Fig. 36 Schematic view of kinetically aggregated filler clusters in rubber below and above the gel point Φ^*: the left side characterizes the local structure of carbon black clusters, build by primary particles and primary aggregates; accordingly, every *black disc* on the right side represents a primary aggregate

5.1
Structure and Elasticity of Filler Networks at Small Strain

5.1.1
Cluster-Cluster Aggregation (CCA) in Elastomers

The kinetic cluster-cluster-aggregation (CCA)-model is based upon the assumption that nanoscopic filler particles, like carbon blacks, are not fixed in space but can perform random movements, i.e., the particles are allowed to fluctuate around their mean position in a polymer matrix. Upon contact of neighboring particles or clusters they stick together, irreversibly. This refers to the fact that the thermal energy of colloidal particles is in general much smaller than their interaction energy. Due to the high viscosity of the polymer, the mobility of the particles is restricted, but it increases significantly with decreasing particle size and increasing temperature because the viscosity decreases. For sufficient high temperature, small filler particles and high filler concentrations, above the gel point ($\Phi>\Phi^*$), the mean distance of the particles becomes smaller than the fluctuation length, implying that the aggregation mechanism is no longer influenced by the restricted mobility of the particles. Under this condition, diffusion limited cluster by cluster aggregation leads to a space-filling configuration of CCA-clusters, similar to colloid aggregation in low viscosity media [21–23, 25, 85]. A schematic view of this situation above the gel point $\Phi>\Phi^*$ is shown in Fig. 36. For filler concentrations just below the gel point, an irregular configuration of partly separated clusters results as shown in the center of Fig. 36. Note that the global connectivity of such systems, as experienced, e.g., by the electrical conductivity measurements in Sect. 4.2, is similar to a percolation structure, since the clusters are "randomly connected" by small conducting gaps. This reflects the fact that the mechanical gel point is generally larger than the electrical percolation threshold ($\Phi^*>\Phi_c$). A discussion of this point is found in [22].

Due to the characteristic self-similar structure of the CCA-clusters with fractal dimension $d_f \approx 1.8$ [3–8, 12], the cluster growth in a space-filling configuration above the gel point Φ^* is limited by the solid fraction Φ_A of the clusters. The cluster size is determined by a space-filling condition, stating that, up to a geometrical factor, the local solid fraction Φ_A equals the overall solid concentration Φ:

$$\Phi_A(\Phi) = N_F^{-1}\Phi \text{ for } \Phi > \Phi* \tag{21}$$

The proportionality constant N_F in Eq. (21) is a generalized Flory-Number of order one ($N_F \cong 1$) that considers a possible interpenetrating of neighboring clusters [22]. For an estimation of cluster size in dependence of filler concentration we take into account that the solid fraction of fractal CCA-clusters fulfils a scaling law similar to Eq. (14). It follow directly from the definition of the mass fractal dimension d_f given by $N_A \cong (\xi/d)^{d_f}$, which implies

$$\Phi_A(\xi) \equiv \frac{V_A}{(\pi/6)\xi^3} = \frac{N_A(\xi)\, d^3 \Phi_p}{\xi^3} \cong \left(\frac{d}{\xi}\right)^{3-d_f} \Phi_p \tag{22}$$

Here, V_A is the solid volume and N_A is the number of particles or primary aggregates of size d in the clusters of size ξ. Φ_p is the solid fraction of primary aggregates considered in Sect. 3.2.2. For spherical filler particles it equals $\Phi_p = 1$.

From a combination of Eqs. (21) and (22) one finds that the cluster size ξ decreases with increasing filler concentration Φ according to a power law. This reflects the fact that smaller clusters occupy less empty space than larger clusters (space-filling condition). It means that the correlation length of the CCA filler network, i.e., the size of the fractal heterogeneity, decreases with increasing filler concentration. This is similar as in percolation theory, though the scaling behavior is generally different.

5.1.2
Elasticity of Flexible Chains of Filler Particles

In view of a micro-mechanical characterization of the elasticity of fractal CCA-filler networks it is sufficient to consider the elastic properties of a single unit cell, because the system is homogeneous on length scales larger than the correlation length ξ. This implies that the elastic modulus G_A of a single CCA-cluster is representative for the whole network. It can be evaluated by referring to Kantor and Webman's model of flexible chains of filler particles that is based on a vectorial Born-lattice model with a tension- and bending energy term between contacting particles [155]. Thereby, we use an approximation of the CCA-cluster backbone as a single spanning arm, i.e., we describe it as a tender, curved rod [21–23]. This is possible, because the CCA-cluster backbone has almost no branches [3, 12], implying that the energy of a strained cluster is primary stored in filler-filler bonds along the connecting

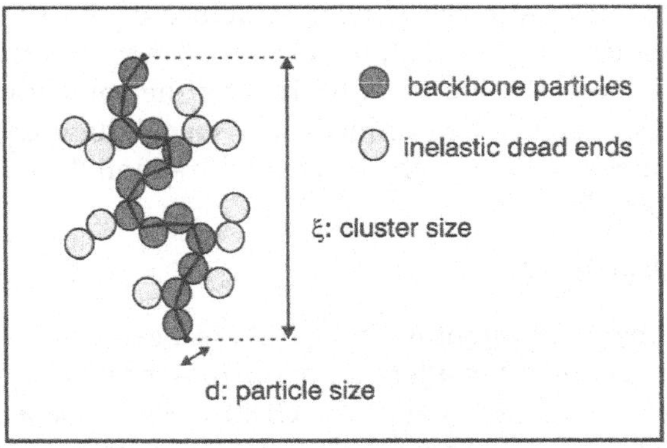

Fig. 37 Schematic view of the decomposition of a CCA-cluster in elastically active backbone particles and inelastic dead ends. Energy is stored along the connecting path of the backbone particles, indicated as the *curved line*

path between the backbone particles. A schematic representation of this approach is depicted in Fig. 37.

According to this model, the clusters act as molecular springs with end to end distance ξ, consisting of N_B backbone units of length d. The connectivity of the backbone units is characterized by the backbone fractal dimension $d_{f,B}$. Due to the fractal nature of CCA-clusters it holds that

$$N_B \cong \left(\frac{\xi}{d}\right)^{d_{f,B}} \tag{23}$$

In the present approximation, $d_{f,B}$ is identified with the minimum fractal dimension, i.e., $d_{f,B}=d_{min}\approx 1.3$ for CCA-clusters [3, 12].

Following the line of Kantor and Webman's two-dimensional model, the strain energy H of a chain composed of a set of N_B singly connected units $\{\boldsymbol{b}_i\}$ of length d under an applied force $\boldsymbol{F}$ at the two ends of the chain is given by [155]

$$H = \frac{F^2 N_B S_\perp^2}{2\,G} + \frac{F^2 L_{II}\, d}{2\,Q} \tag{24}$$

where

$$S_\perp^2 = \frac{1}{F^2 N_B} \sum_{i=1}^{N_B} \left[(\boldsymbol{F} \times \boldsymbol{z})\,(\boldsymbol{R}_{i-1} - \boldsymbol{R}_{N_B})\right]^2 \tag{25}$$

is the squared radius of gyration of the projection of the chain on a two dimensional plane and

$$L_{II} = \frac{1}{F^2 d} \sum_{i=1}^{N_B} (\boldsymbol{F} \cdot \boldsymbol{b}_i)^2 \tag{26}$$

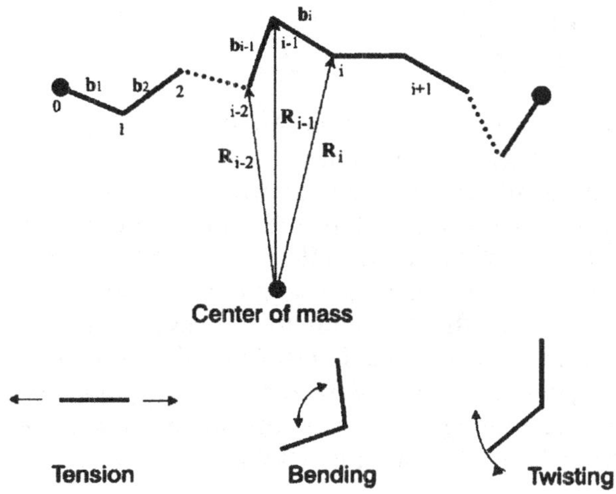

Fig. 38 Illustration of Kantor and Webman's model of flexible chains with tension-, bending-, and twisting energy terms [155]

Here, G and Q are local elastic constants corresponding to the changes of angles between singly connected bonds and longitudinal deformation of single bonds, respectively. The vector z is a unit vector perpendicular to the plane. For long chains the second term in Eq. (24) can be neglected and the major part of the strain energy H results from the first term, i.e., the bending term of the chain. Then, the force constant k of the chain, relating the elastic energy to the displacement squared of the end of the chain, is well approximated by the bending force constant k_S [155]:

$$k \approx k_S = G/(N_B S_{\perp}^2) \tag{27}$$

The mathematical treatment of a three-dimensional generalization of this linear elastic model is more complex. In this case the angular deformation is not limited to the on-plane bending, but also off-plane twisting takes place. This model is depicted schematically in Fig. 38.

In a simplified approach introduced by Lin and Lee [56], the contributions from the two different kinds of angular deformation, bending and twisting, have been taken into account by an averaged bending-twisting deformation. This is obtained by replacing the elastic constant G, in the first term of Eq. (24), through an averaged elastic constant $\overline{G}$ of different kinds of angular deformations. Then, by using the relation $S_{\perp} \cong \xi$ and Eq. (23), one obtains for the force constant $k \approx k_S$ of the cluster backbone:

$$k_S = \frac{\kappa \overline{G}}{d^2} \left(\frac{d}{\xi}\right)^{2+d_{f,B}} \tag{28}$$

where κ is a geometrical factor of order one. The elastic modulus of the cluster backbone is then found as

$$G_A \equiv \xi^{-1} k_S = \frac{\kappa \overline{G}}{d^3} \left(\frac{d}{\xi} \right)^{3+d_{f,B}} \tag{29}$$

Equation (29) describes the modulus G_A of the clusters as a local elastic bending-twisting energy term $\overline{G}$ times a scaling function that involves the size and geometrical structure of the clusters. Note that in the case of a linear cluster backbone with $d_{f,B}=1$, Eqs. (28) and (29) correspond to the well know elastic behavior of linear, flexible rods, where the bending modulus falls off with the fourth power of the length ξ. The above approach represents a generalization of this behavior to the case of curved, flexible rods. Experimental evidence for the elastic response of filler clusters of nano-particle chain aggregates is given by Friedlander et al. [156, 157], who showed by TEM analysis that aggregates of inorganic metal oxides stretch under tension and contract when the tension is relaxed.

5.1.3
Scaling Behavior of the Small Strain Modulus

The small strain viscoelastic behavior of filler reinforced rubbers is well known to be strongly affected by the properties of the filler network. This refers to the fact that the elastic modulus G_A of the filler network units, i.e., the CCA clusters, is generally much larger than the elastic modulus G_R of the rubber and a rigidity condition $G_A >> G_R$ is fulfilled. However, from Eq. (29) it is clear that the rigidity condition cannot be fulfilled in all cases, because the modulus G_A of the clusters decreases rapidly with increasing size of the clusters. It means that only relatively small filler clusters can reinforce a rubber matrix with $G_R \cong 0.1$ MPa, since large clusters are too soft. In the present linear viscoelastic model of the small strain behavior of elastomer composites, only this case is considered, i.e., the rigidity condition $G_A >> G_R$ is assumed to be fulfilled.

For filler concentrations above the gel point Φ^*, where a through-going filler network is formed, stress between the (closely packed) CCA-clusters is transmitted directly between the spanning arms of the clusters that bend substantially. In this case, the strain of the rubber is almost equal to the strain of the spanning arms of the clusters ($\varepsilon_R \approx \varepsilon_A$). It means that, due to the rigidity condition, $G_A >> G_R$, the overwhelming part of the elastic energy is stored in the bending arms of the clusters and the contribution of the rubber to the small strain modulus G'_0 of the sample can be neglected, i.e., $G'_0 \approx G_A$. Accordingly, the stored energy density (per unit strain) of highly filled elastomers can be approximated by that of the filler network. This in turn equals the stored energy density of a single CCA-cluster, due to the homogeneity of the filler network on length scales above the cluster size ξ.

In the framework of the approximation given by the rigidity condition, a simple power law relation can be derived for the dependency of the small strain modulus G'_0 of the composite on filler concentration Φ. It is obtained,

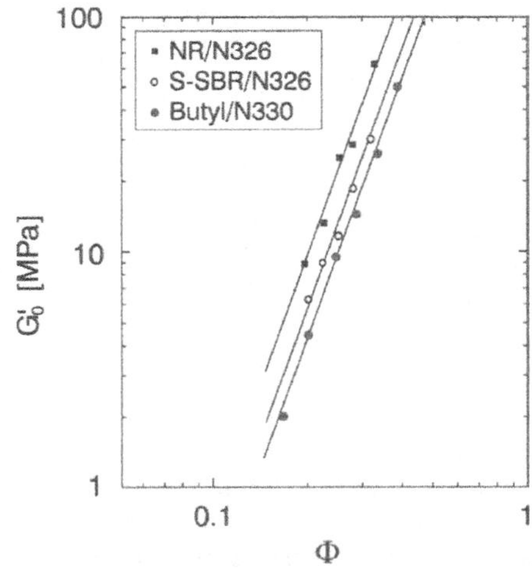

Fig. 39 Double logarithmic plot of the small strain storage modulus vs filler volume fraction for a variety of carbon black filled composites, as indicated. The *solid lines* with slope 3.5 correspond to the prediction of Eq. (30). Data of butyl/N330 composites are taken from [28] (compare Fig. 2a)

if G_A in Eq. (29) is expressed by the solid fraction Eq. (22) and the space-filling condition Eq. (21) is used:

$$G'_o \cong \frac{\overline{G}}{d^3}\left(\frac{\Phi_A}{\Phi_p}\right)^{(3+d_{f,B})/(3-d_f)} \cong \frac{\overline{G}}{d^3}\left(\frac{\Phi}{\Phi_p}\right)^{(3+d_{f,B})/(3-d_f)} \tag{30}$$

Accordingly, we expect a power law behavior $G'_0 \sim (\Phi/\Phi_p)^{3.5}$ of the small strain elastic modulus for $\Phi > \Phi^*$. Thereby, the exponent $(3+d_{f,B})/(3-d_f) \approx 3.5$ reflects the characteristic structure of the fractal heterogeneity of the filler network, i.e., the CCA-clusters. The strong dependency of G'_0 on the solid fraction Φ_p of primary aggregates reflects the effect of "structure" on the storage modulus.

The predicted power law behavior $G'_0 \sim \Phi^{3.5}$ for filler concentrations above the gel point is confirmed by the experimental results depicted in Fig. 39 and Fig. 40, where the small strain storage modulus of a variety of filled rubbers is plotted against filler loading in a double logarithmic manner. Figure 39 shows the scaling properties for carbon black composites, i.e., for the low structure black N326 in NR and S-SBR, respectively, and the classical butyl/N330-data of Payne [28], already depicted in Fig. 2a. Within the framework of experimental errors, Eq. (30) is found to be fairly well fulfilled for all systems. Note however that the 3.5-power law is in general not fulfilled for high structure blacks, since for such systems Φ_p cannot be considered to be independent of filler concentration, but increases due to an enhanced aggregate fracture with increasing filler loading (compare Sect. 3.2.2).

Figure 40 demonstrates that the 3.5-power law also holds for NR composites with inorganic- and polymeric fillers, respectively. Beside a technical sil-

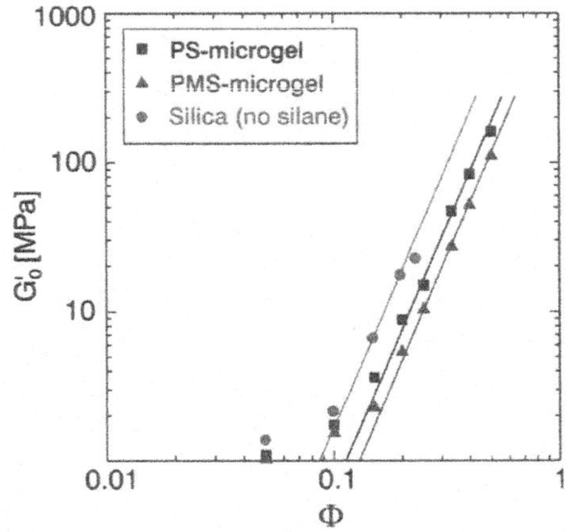

Fig. 40 Double logarithmic plot of the small strain storage modulus vs filler volume fraction for a variety of NR composites, as indicated. The *solid lines* with slope 3.5 correspond to the prediction of Eq. (30) [158, 159]

ica filler, polystyrene (PS)- and poly-methoxy-styrene (PMS) microgels are applied as model fillers of definite size (d=100 nm) with a very narrow distribution width. Since these model fillers consist of spherical particles with Φ_p=1, no disturbing influence of mixing severity on Φ_p, as found, e.g., for the SBR/N339 composites in Sect. 3.2.2, can appear. The scaling factors of the solid lines allow for an approximate estimation of the front factors of Eq. (30). It yields a typical apparent energy density $\bar{G}/d^3$ of filler-filler bonds of the order of 1 GPa, corresponding to an energy $\bar{G} \cong 10^{-12}$J.

The predicted scaling behavior Eq. (30) is also found to be well fulfilled for carbon black suspensions in ethylene-vinyl-acetate copolymers [51]. Furthermore, it is confirmed by viscoelastic data obtained for S-SBR composites with highly cross-linked BR-microgels of various size [57].

Equation (39) describes the small strain modulus G'_0 as product of a local elastic constant$\bar{G}/d^3$ of filler-filler bonds and a geometrical factor, considering the structure of the filler network. Accordingly, the temperature- or frequency dependency of G'_0 is determined by the bending-twisting energy $\bar{G}$ of the filler-filler bonds, which is controlled by the bound rubber phase around the filler clusters. Coming back to the model of filler-filler bonds developed in Sect. 4.3 and depicted in Fig. 33, the energy in a strained bond is primary stored in a nanoscopic bridge of immobilized, glassy polymer between the filler particles, implying that the temperature- or frequency dependence of G'_0 is given by that of the glassy polymer. Consequently, for highly filled rubbers above the glass transition temperature, we expect an Arrhenius temperature behavior typical for polymers in the glassy state. This is in agreement with experimental findings [26–28].

We finally note that the decomposition of G'_0 into a local elastic constant and a geometrical factor implies that the same form as Eq. (30) must also hold for the loss modulus G''_0. This follows from the fact that in the linear viscoelastic regime any geometrical factor must act on the real and imaginary part of the complex modulus in the same way. Otherwise, the Kramers-Kronig relations, indicative for linear systems, cannot be fulfilled [160]. From experimental data this is not always obvious, since the relatively small phase angle at small strain amplitudes can not be estimated as accurate as the norm of the modulus. In particular, the viscoelastic data of Payne [28] on butyl/N330-composites, depicted in Fig. 2b, show that the tan δ at small strain is almost independent of filler loading, implying that in the linear regime storage and loss modulus transform with the same geometrical factor.

5.2 Stress Softening and Filler-Induced Hysteresis at Large Strain

5.2.1 *Strength and Fracture of Filler Clusters in Elastomers*

So far, we have considered the elasticity of filler networks in elastomers and its reinforcing action at small strain amplitudes, where no fracture of filler-filler bonds appears. With increasing strain, a successive breakdown of the filler network takes place and the elastic modulus decreases rapidly if a critical strain amplitude is exceeded (Fig. 42). For a theoretical description of this behavior, the ultimate properties and fracture mechanics of CCA-filler clusters in elastomers have to be evaluated. This will be a basic tool for a quantitative understanding of stress softening phenomena and the role of fillers in internal friction of reinforced rubbers.

The failure- or yield strain ε_F of filler clusters in elastomers can be estimated by referring to the elasticity model of flexible chains with tension- and bending-twisting energy terms, introduced above. According to this model, a single cluster corresponds to a series of two molecular springs: a soft one, representing the bending-twisting mode, and a stiff one, representing the tension mode. This mechanical equivalence is illustrated in Fig. 41. On the one hand, the soft spring with force constant $k_S \sim \bar{G}$ governs the elasticity of the whole system, provided that the deformation of the stiff spring can be neglected. On the other hand, the stiff spring impacts the fracture behavior of the system, since it considers the longitudinal deformation and, hence, separation of filler-filler bonds. The cluster breaks down when a critical separation of bonded filler particles is exceeded, i.e., if the failure strain ε_b of filler-filler bonds is reached. The failure strain ε_F of the filler cluster is then determined by the stress equilibrium between the two springs, which reads in the case of large clusters with $k_b >> k_S$:

$$\varepsilon_F = \left(1 + \frac{k_b}{k_s}\right)\varepsilon_b \approx \frac{Q\varepsilon_b}{\kappa\,\overline{G}}\left(\frac{\xi}{d}\right)^{2+d_{f,B}} \tag{31}$$

Here, $k_b = Q/d^2$ is the force constant of longitudinal deformations of filler-filler bonds and k_S is the bending-twisting force constant of the cluster, which is given by Eq. (28). From Eq. (31) one finds that the yield strain of a

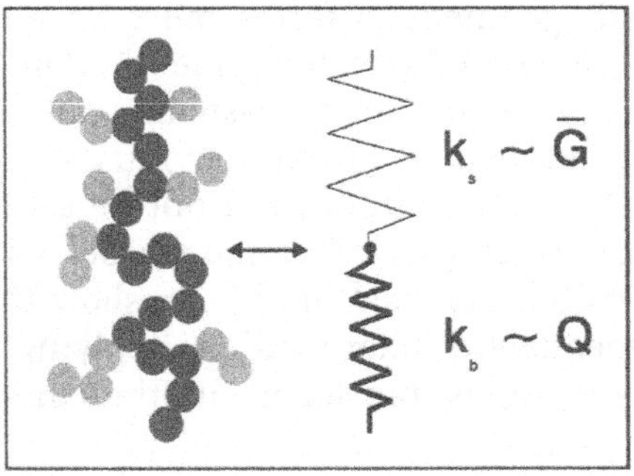

Fig. 41 Schematic view demonstrating the mechanical equivalence between a filler cluster and a series of soft and stiff molecular springs, representing bending-twisting- and tension deformation of filler-filler bonds, respectively

filler cluster depends on the ratio between the elastic constants $Q/\bar{G}$ and increases with the cluster size ξ according to a power law. Consequently, larger clusters show a higher extensibility than smaller ones, due to the ability to bend and twist around the bonds. This kind of elastic behavior of filler clusters and the dependence of strength on cluster size implies a high flexibility of filler clusters in strained rubbers. It plays a crucial role in rubber reinforcement even up to large strains. In particular, the Payne effect, shown, e.g., in Fig. 2 and Fig. 42, can be modeled by referring to a specific size distribution of clusters in the filler network and assuming a successive breakdown of filler clusters with increasing strain [56]. Then, the broadness of the transition regime of the storage modulus (about three decades of strain) reflects the width of the size distribution of clusters in the filler network.

Equations (29)–(31) allow for a qualitative analysis of the transition of the elastic modulus from the small strain plateau value to the high strain non-linear behavior. This transition becomes apparent in Fig. 42, showing the strain dependent storage modulus of EPDM- and S-SBR composites filled with the same amount of graphitized and non-graphitized N220, respectively. Since in first approximation the morphology of primary carbon black aggregates can be considered to be not much different, i.e., d=constant and Φ_p=constant, Eqs. (29) and (30) imply with $G_A \approx G'_0$ that the cluster size ξ is roughly the same for all four systems. Hence, the differences observed in the critical strain amplitude ε_{app}, where non-linearity appears, and those in the small strain modulus G'_0 have to be related to differences in the elastic constants Q and $\bar{G}$. Due to Eq. (29), the higher values of G'_0 for the EPDM-composites and the systems with the graphitized black, respectively, indicate stiffer filler-filler bonds with larger values of the bending-twisting constant $\bar{G}$. Equation (31) implies that the lower values of ε_{app} for the systems with the graphitized black can partly be related to the same effect, i.e., larger values of $\bar{G}$, provided ε_{app}~ε_F is assumed. However, a closer analysis shows that the shift factor describing the variation of ε_{app} is significantly larger than that of G'_0. This emphasizes that the systems with the graphitized black ex-

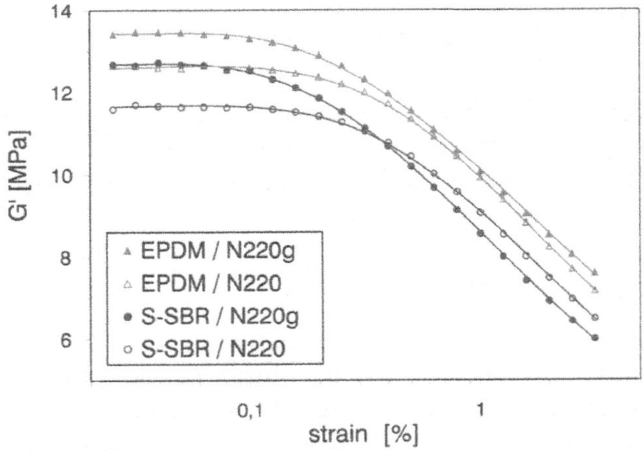

Fig. 42 Strain dependency of the storage modulus at 20 °C of cross-linked S-SBR- and EPDM composites filled with 50 phr N220 and graphitized N220g, respectively

hibit smaller values of the tension energy Q of filler-filler bonds, reflecting a lower interaction strength between the particles.

These results can be well understood, if the model of filler-filler bonds developed in Sect. 4.3 is considered, again. It implies that the local elastic constants Q and $\bar{G}$ are controlled by the rubber phase around fillers, i.e., they are attributed to the amount of bound rubber and its influence on the spacing of contact gaps. The elastic constant Q is also affected by the van der Waals force between contacting filler particles, representing the driving force of filler aggregation. A comparison of the S-SBR-systems shown in Fig. 42 and the annealed systems in Fig. 35 makes clear that cross-linking has no pronounced effect on the viscoelastic data, though the difference between both filler types decreases during cross-linking and the cross-over point is shifted to somewhat smaller strain amplitudes. Note that, apart from cross-linking, the systems have been prepared with respect to the same procedure, but the measurement temperature is different.

For analyzing the fracture behavior of filler clusters in strained rubbers, it is necessary to estimate the strain of the clusters in dependence of the external strain of the samples. In the case of small strains, considered above, both strain amplitudes in spatial direction μ are equal ($\varepsilon_{A,\mu}=\varepsilon_\mu$), because the stress is transmitted directly between neighboring clusters of the filler network. For strain amplitudes larger than about 1%, this is no longer the case, since a gel-sol transition of the filler network takes place with increasing strain [57, 154] and the stress of the filler clusters is transmitted by the rubber matrix. At larger strains, the local strain $\varepsilon_{A,\mu}$ of a filler cluster in a strained rubber matrix can be determined with respect to the external strain ε_μ, if a stress equilibrium between the strained cluster and the rubber matrix is assumed $\left(\varepsilon_{A,\mu}G_A(\xi_\mu)=\hat{\sigma}_{R,\mu}(\varepsilon_\mu)\right)$. With Eq. (29) this implies

$$\varepsilon_{A,\mu}(\varepsilon_\mu) = \frac{d^3}{\kappa G}\left(\frac{\xi_\mu}{d}\right)^{3+d_{f,B}} \hat{\sigma}_{R,\mu}(\varepsilon_\mu) \tag{32}$$

Here, ξ_μ denotes the cluster size in spatial direction μ of the main axis system and $\hat{\sigma}_{R,\mu}(\varepsilon_\mu)$ is the norm of the relative stress of the rubber with respect to the initial stress at the beginning of each strain cycle, where $\partial\varepsilon_\mu/\partial t=0$:

$$\hat{\sigma}_{R,\mu}(\varepsilon_\mu) \equiv \left|\sigma_{R,\mu}(\varepsilon_\mu) - \sigma_{R,\mu}(\partial\varepsilon_\mu/\partial t = 0)\right| \tag{33}$$

The application of this normalized, relative stress in Eq. (32) is essential for a constitutive formulation of cyclic cluster breakdown and re-aggregation during stress-strain cycles. It implies that the clusters are stretched in spatial directions with $\partial\varepsilon_\mu/\partial t>0$, only, since $\varepsilon_{A,\mu}\geq 0$ holds due to the norm in Eq. (33). In the compression directions with $\partial\varepsilon_\mu/\partial t<0$ re-aggregation of the filler particles takes place and the clusters are not deformed. An analytical model for the large strain non-linear behavior of the nominal stress $\sigma_{R,\mu}(\varepsilon_\mu)$ of the rubber matrix will be considered in the next section.

A comparison of Eqs. (31) and (32) makes clear that for large deformations, when the stress of the clusters is transmitted by the rubber matrix, the strain $\varepsilon_{A,\mu}$ of the clusters increases faster with their size ξ_μ than the failure strain $\varepsilon_{F,\mu}$. Accordingly, with increasing strain the large clusters in the system break first followed by the smaller ones. The maximum size ξ_μ of clusters surviving at exposed external strain ε_μ is estimated by the stress equilibrium between the rubber matrix and the failure stress $\sigma_{F,\mu}=\varepsilon_{F,\mu}G_A(\xi_\mu)$ of the clusters:

$$\xi_\mu(\varepsilon_\mu) = \frac{Q\varepsilon_b}{d^2\hat{\sigma}_{R,\mu}(\varepsilon_\mu)} \tag{34}$$

This allows for an evaluation of the stress contribution of the stretched filler clusters if the size distribution of the clusters in the system is known. Note that for small deformations, where $\varepsilon_{A,\mu}=\varepsilon_\mu$ holds, the situation is different, since $\varepsilon_{A,\mu}$ is independent of cluster size. Then, the small clusters break first followed by the larger ones.

5.2.2 *Free Energy Density of Reinforced Rubbers*

Starting from a particular size distribution of kinetically aggregated filler clusters in a rubber matrix, we can now formulate the free energy density of a highly strained sample by assuming the following microscopic scenario. With increasing strain of a virgin sample, a successive breakdown of filler clusters takes place under the exposed stress of the bulk rubber. This process begins with the largest clusters and continues up to a minimum cluster size $\xi_{\mu,\min}$ in spatial direction μ, which is given by Eq. (34) evaluated for the maximum stress of the rubber matrix $\hat{\sigma}_{R,\mu}(\varepsilon_{\mu,\max})$ reached at maximum ex-

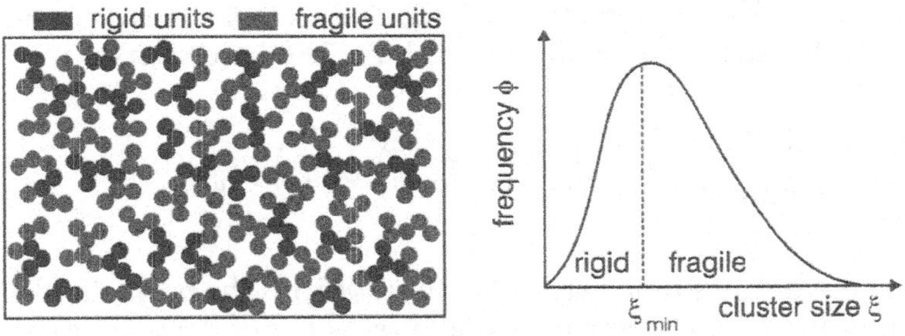

Fig. 43 Schematic view of the decomposition of filler clusters in rigid and fragile units for pre-conditioned samples. The *right side* shows the cluster size distribution with the pre-strain dependent boundary size ξ_{min}

ternal strain $\varepsilon_{\mu,max}$. During the back-cycle, complete re-aggregation takes place, but the filler-filler bonds that are formed after once being broken are significantly weaker and softer than the original annealed ones. For subsequent stress-strain cycles of a pre-conditioned sample, two micro-mechanical mechanisms of rubber reinforcement are distinguished:

1. Hydrodynamic reinforcement of the rubber matrix by the fraction of hard, rigid filler clusters with strong filler-filler bonds that have not been broken during previous deformations.
2. Cyclic breakdown and re-aggregation of the residual fraction of fragile, soft filler clusters with weaker filler-filler bonds.

The fraction of rigid filler clusters decreases with increasing pre-strain, while the fraction of fragile filler clusters increases. The decomposition into rigid and fragile filler cluster units is illustrated in Fig. 43. The fragile filler clusters with soft bonds bend substantially in the stress field of the rubber, implying that their contribution to hydrodynamic reinforcement is relatively small. The mechanical action of the fragile filler clusters refers primary to a viscoelastic effect, since any type of cluster that is stretched in the stress field of the rubber stores energy that is dissipated when the cluster breaks. This mechanism leads to a filler-induced viscoelastic contribution to the total stress that impacts the internal friction of filled rubber samples, significantly. Note that this kind of viscoelastic response is present also in the limit of quasi-static deformations, where no explicit time dependency of the stress-strain cycles is taken into account.

According to these considerations, we assume that for quasi-static, cyclic deformations of filler reinforced rubbers up to large strain the total free energy density consists of two contributions:

$$W(\varepsilon_\mu) = W_R(\varepsilon_\mu) + W_A(\varepsilon_\mu) \qquad (35)$$

The first addend is the equilibrium energy density stored in the extensively strained rubber matrix, which includes hydrodynamic reinforcement by a fraction of rigid filler clusters (see below). The second addend considers the energy stored in the substantially strained fragile filler clusters:

$$W_A(\varepsilon_\mu) = \sum_{\mu}^{\dot{\varepsilon}_\mu > 0} \frac{1}{2d} \int_{\xi_{\mu,\min}}^{\xi_\mu(\varepsilon_\mu)} G_A\left(\xi'_\mu\right) \varepsilon^2_{A,\mu}\left(\xi'_\mu, \varepsilon_\mu\right) \phi\left(\xi'_\mu\right) d\xi'_\mu \tag{36}$$

where the dot denotes time derivative. Here, $\phi(\xi_\mu)$ is the normalized size distribution of the CCA-clusters in spatial direction μ of the main axis system. It can be evaluated by referring to Smoluchowski's equation for the kinetics of irreversible cluster-cluster aggregation of colloids [8, 161, 162]. In the case of large clusters (and large times), one can obtain a reduced form of the size distribution function [8]:

$$\phi(\xi_\mu) = \frac{4d}{<\xi_\mu>} \left(\frac{\xi_\mu}{<\xi_\mu>}\right)^{-2\Omega} exp - \left(\frac{(1-2\Omega)\,\xi_\mu}{<\xi_\mu>}\right) \quad \text{for} \quad \mu = 1,2,3 \tag{37}$$

For $\Omega<0$, this distribution function is peaked around a maximum cluster size $(2\Omega/(2\Omega-1))<\xi_\mu>$, where $<\xi_\mu>$ is the mean cluster size. $2\Omega=\alpha+d_f^{-1}$ is a parameter describing details of the aggregation mechanism, where α' is an exponent considering the dependency of the diffusion constant Δ of the clusters on its particle number, i.e., $\Delta\sim N_A^{\alpha'}$. This exponent is in general not very well known. In a simple approach, the particles in the cluster can assumed to diffusion independent from each other, as, e.g., in the Rouse model of linear polymer chains. Then, the diffusion constant varies inversely with the number of particles in the cluster ($\Delta\sim N_A^{-1}$), implying $2\Omega=-0.44$ for CCA-clusters with characteristic fractal dimension $d_f=1.8$.

The sum in Eq. (36) is taken over the stretching directions with $\partial\varepsilon_\mu/\partial t>0$, only, implying that the free energy density is anisotropic. This corresponds to the assumption that clusters are strained and successively broken in stretching directions, while healing of the clusters takes place in the compression directions. It insures that a cyclic breakdown and re-aggregation of clusters can be described. The integration in Eq. (36) is performed over the fraction of fragile filler clusters that are not broken at exposed strain ε_μ of the actual cycle (Eq. 34). The clusters smaller than $\xi_{\mu,\min}$, representing the fraction that survived the first deformation cycle, are not considered in Eq. (36). Due to the stiff nature of their filler-filler bonds, referring to the bonds in the virgin state of the sample, these clusters behave rigid and give no contribution to the stored energy. Instead, they dominate the hydrodynamic reinforcement of the rubber matrix. This is specified by the strain amplification factor X, which relates the external strain ε_μ of the sample to the internal strain ratio λ_μ of the rubber matrix:

$$\lambda_\mu = 1 + X\varepsilon_\mu \tag{38}$$

In the case of a pre-conditioned sample and for strains smaller than the previous straining ($\varepsilon_\mu < \varepsilon_{\mu,\max}$), the strain amplification factor is constant and determined by $\varepsilon_{\mu,\max}$ ($X=X(\varepsilon_{\mu,\max})$). For the first deformation of virgin samples it depends on the external strain ($X=X(\varepsilon_\mu)$). By applying a relation derived by Huber and Vilgis [64, 65] for the strain amplification factor of overlapping fractal clusters, $X(\varepsilon_{\mu,\max})$ or $X(\varepsilon_\mu)$ can be evaluated by averaging over the size distribution of rigid clusters in all space directions. In the case of pre-conditioned samples this yields

$$X\left(\varepsilon_{\mu,\max}\right) = 1 + c\left(\frac{\Phi}{\Phi_p}\right)^{\frac{2}{3-d_f}} \sum_{\mu=1}^{3} \frac{1}{d} \int_0^{\xi_{\mu,\min}} \left(\frac{\xi'_\mu}{d}\right)^{d_w - d_f} \phi\left(\xi'_\mu\right) d\xi'_\mu \tag{39}$$

where c is a constant of order one and $d_w \approx 3.1$ [23] is the anomalous diffusion exponent on fractal CCA-clusters. For the deformation of a virgin sample, $X(\varepsilon_\mu)$ is obtained in a similar way by performing the integration from zero up to the strain dependent cluster size $\xi_\mu(\varepsilon_\mu)$ determined by Eq. (34).

An important role in the present model is played by the strongly non-linear elastic response of the rubber matrix that transmits the stress between the filler clusters. We refer here to an extended tube model of rubber elasticity, which is based on the following fundamental assumptions. The network chains in a highly entangled polymer network are heavily restricted in their fluctuations due to packing effects. This restriction is described by virtual tubes around the network chains that hinder the fluctuation. When the network elongates, these tubes deform non-affinely with a deformation exponent $\nu=1/2$. The tube radius r_μ in spatial direction μ of the main axis system depends on the deformation ratio λ_μ as follows:

$$r_\mu = r_0 \lambda_\mu^\nu \tag{40}$$

where r_0 is the tube radius in the non-deformed state. The assumption of the non-affine tube deformation ($\nu=1/2$) is essential. It was initially derived based upon fundamental molecular statistical calculations [24, 163, 164] and later on confirmed by applying scaling arguments [77, 78, 165]. Experimental evidence of non-affine tube deformations according to Eq. (40) is provided by neutron scattering of strained rubbers [166] as well as stress-strain measurements on swollen polymer networks [167].

An extension of the non-affine tube model for applications up to large strains is obtained by considering that the network chains have a finite length and the stress in the network becomes infinitely large, when the chain sections between two subsequent trapped entanglements are stretched fully. The trapping of chain entanglements by two cross-link points prevents the sliding of the chains across each other under extension, implying that the entanglement becomes an elastically effective network junction. The free energy density of the extended, non-Gaussian tube model with non-affine tube deformation is then as follows [75–78, 168]:

$$W_R(\varepsilon_\mu) = \frac{G_c}{2}\left\{\frac{\left(\sum\limits_{\mu=1}^{3}\lambda_\mu^2 - 3\right)\left(1 - \frac{T_e}{n_e}\right)}{1 - \frac{T_e}{n_e}\left(\sum\limits_{\mu=1}^{3}\lambda_\mu^2 - 3\right)} + \ln\left[1 - \frac{T_e}{n_e}\left(\sum_{\mu=1}^{3}\lambda_\mu^2 - 3\right)\right]\right\}$$
$$+2G_e\left(\sum_{\mu=1}^{3}\lambda_\mu^{-1} - 3\right) \tag{41}$$

Here, n_e is the number of statistical chain segments between two successive entanglements and T_e is the trapping factor ($0<T_e<1$), which characterizes the portion of elastically active entanglements. The first bracket term of Eq. (41) considers the constraints due to inter-chain junctions, with an elastic modulus G_c proportional to the density of network junctions. The second addend is the result of tube constraints, whereby G_e is proportional to the entanglement density μ_e of the rubber. The parenthetical expression in the first addend takes into account the finite chain extensibility by referring to a proposal of Edwards and Vilgis [169]. For the limiting case $n_e/T_e=\sum\lambda_\mu^2-3$, a singularity is obtained for W_R. This happens when the chains between successive trapped entanglements are fully stretched out. It makes clear that the approach in Eq. (41) characterizes trapped entanglements as some kind of physical cross-links (slip-links) that dominate the extensibility of the network due to the larger number of entanglements as compared to chemical cross-links. In the limit $n_e\rightarrow\infty$ the original Gaussian formulation of the non-affine tube model, derived by Heinrich et al. [24] for infinite long chains, is recovered.

The trapping factor T_e increases as the cross-link density increases, whereas n_e and G_e—as terms that are specific to the polymer—are to a great extent independent of cross-link density. For the cross-link and tube constraint moduli, the following relations to molecular network parameters hold:

$$G_c = A_c \nu_{mech} k_B T \tag{42}$$

$$G_e = \frac{\rho N_a l_s^2 k_B T}{4\sqrt{6} M_s r_o^2} \tag{43}$$

Here, ν_{mech} is the mechanically effective chain density specified, e.g., in [168], $A_c\approx 0.67$ [170] is a microstructure factor which describes the fluctuations of network junctions, N_a the Avogadro number, ρ mass density, M_s and l_s molar mass and length of a statistic segment, respectively, k_B the Boltzmann constant, and T absolute temperature.

From Eq. (41) the nominal stress $\sigma_{R,\mu}$ that relates the force F_μ in spatial direction μ to the initial cross section $A_{o,\mu}$ is found by differentiation, $\sigma_{R,\mu}=\partial W_R/\partial\lambda_\mu$. For uniaxial extensions of unfilled rubbers ($X=1$) with $\lambda_1=\lambda$, $\lambda_2=\lambda_3=\lambda^{-1/2}$ the following relation can be derived:

$$\sigma_{R,1} = G_c\left(\lambda - \lambda^{-2}\right)\left\{\frac{1 - T_e/n_e}{\left(1 - \frac{T_e}{n_e}\left(\lambda^2 + 2/\lambda - 3\right)\right)^2} - \right.$$

$$\left. - \frac{T_e/n_e}{1 - \frac{T_e}{n_e}\left(\lambda^2 + 2/\lambda - 3\right)}\right\} + 2G_e\left(\lambda^{-1/2} - \lambda^{-2}\right) \tag{44}$$

For equi-biaxial extensions with $\lambda_1=\lambda_2=\lambda$, $\lambda_3=\lambda^{-2}$ one finds for the nominal stress:

$$\sigma_{R,1} = G_c\left(\lambda - \lambda^{-5}\right)\left\{\frac{1 - T_e/n_e}{\left(1 - \frac{T_e}{n_e}\left(2\lambda^2 + \lambda^{-4} - 3\right)\right)^2} - \right.$$

$$\left. - \frac{T_e/n_e}{1 - \frac{T_e}{n_e}\left(2\lambda^2 + \lambda^{-4} - 3\right)}\right\} + 2G_e\left(\lambda - \lambda^{-2}\right) \tag{45}$$

In the case of a pure-shear deformation with $\lambda_1=\lambda$, $\lambda_2=1$, and $\lambda_3=\lambda^{-1}$ one obtains

$$\sigma_{R,1} = G_c\left(\lambda - \lambda^{-3}\right)\left\{\frac{1 - T_e/n_e}{\left(1 - \frac{T_e}{n_e}\left(\lambda^2 + \lambda^{-2} - 2\right)\right)^2} - \right.$$

$$\left. - \frac{T_e/n_e}{1 - \frac{T_e}{n_e}\left(\lambda^2 + \lambda^{-2} - 2\right)}\right\} + 2G_e\left(1 - \lambda^{-2}\right) \tag{46}$$

By fitting experimental data for different deformation modes to these functions, the three network parameters of unfilled polymer networks G_c, G_e, and n_e/T_e can be determined. The validity of the concept can be tested if the estimated fitting parameters for the different deformation modes are compared. A "plausibility criterion" for the proposed model is formulated by demanding that all deformation modes can be described by a single set of network parameters. The result of this plausibility test is depicted in Fig. 44, where stress-strain data of an unfilled NR-vulcanizate are shown for the three different deformation modes considered above. Obviously, the material parameters found from the fit to the uniaxial data provide a rather good prediction for the two other modes. The observed deviations are within the range of experimental errors.

It should be pointed out that the material parameter G_e can be determined in principle more precisely by means of equi-biaxial measurements than by uniaxial measurements. This is due to the fact that the first addend of the G_e-term in Eq. (45) increases linearly with λ. This behavior results from the high lateral contraction on the equi-biaxial extension ($\lambda_3=\lambda^{-2}$). It postulates a close dependency of the equi-biaxial stress on the tube constraint modu-

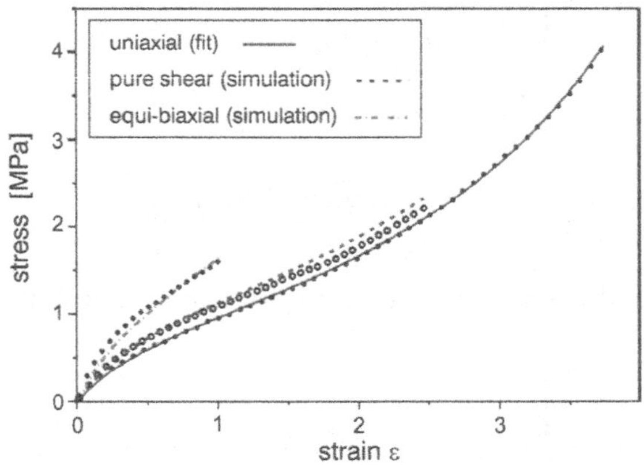

Fig. 44 Stress-strain data (*symbols*) and simulation curves (Eqs. 44–46) of an unfilled NR for three deformation modes. The model parameters are found from a fit to the uniaxial data (G_c=0.43 MPa, G_e=0.2 MPa, n_e/T_e=68)

lus, since G_c and G_e contribute nearly equally to stress, also at larger extensions. For the uniaxial extensions described in Eq. (44) this is not the case. Here, the tube constraints lead to significant effects only in the region of lower extensions, since the G_e-term in Eq. (44) approaches zero as the λ values increase. Nevertheless, the experiments can be carried out more easily in the uniaxial case, and as a result, more reliable experimental data can be obtained.

For practical applications, the parameter G_e can also be determined from the value of the plateau modulus $G_N{}^o$, since the relationship $G_e \approx 1/2 G_N{}^o$ applies in accordance with the tube model. This implies that the parameter G_e is not necessary a fit parameter but rather it is specified by the microstructure of the rubber used. Note that the fit value G_e=0.2 MPa obtained in Fig. 44 is in fair agreement with the above relation, since $G_N{}^o \approx 0.58$ MPa is found for uncross-linked NR-melts [171].

A comparison of the predictions of the extended tube model to stress-strain data of unfilled rubbers offered good agreement for various polymers and cross-linking systems [75–78, 167, 170, 172, 173]. Further confirmation of the non-affine tube approach was obtained in recent investigations considering mechanical stress-strain data and a transversal NMR-relaxation analysis of differently prepared NR-networks [168]. A representation of the molecular statistical foundations of tube-like topological constraints in strained polymer networks is found in [76], where the path integral formulation of rubber elasticity is briefly reviewed. In the second part of [76], a first attempt for considering stress softening phenomena of filled rubbers on the basis of the extended tube model is formulated. Subsequent investigations have shown that a constitutive formulation of stress softening fulfilling the "plausibility criterion" cannot be achieved in this simplified approach [89]. In the following section we will see that the extended model, as developed above, passes the "plausibility test" fairly well also for filled rubbers.

The important point lies in the determination of the topological tube constraint modulus G_e according to its physical value $G_e \approx 1/2 G_N^{\circ}$, which was not realized in the first considerations.

5.2.3 Stress-Strain Cycles of Filled Rubbers in the Quasi-Static Limit

For filler reinforced rubbers, both contributions of the free energy density Eq. (35) have to be considered and the strain amplification factor X, given by Eq. (39) differs from one. The nominal stress contributions of the cluster deformation are determined by $\sigma_{A,\mu} = \partial W_A / \partial \varepsilon_{A,\mu}$, where the sum over all stretching directions, that differ for the up- and down cycle, have to be considered. For uniaxial deformations $\varepsilon_1 = \varepsilon$, $\varepsilon_2 = \varepsilon_3 = (1+\varepsilon)^{-1/2} - 1$ one obtains a positive contribution to the total nominal stress in stretching direction for the up-cycle if Eqs. (29)–(36) are used:

$$\sigma_{0,1}^{up}(\varepsilon) = \sigma_{R,1}(\varepsilon) + \hat{\sigma}_{R,1}(\varepsilon) \int_{\frac{Q\varepsilon_b}{d^3 \hat{\sigma}_{R,1}(\varepsilon_{\max})}}^{\frac{Q\varepsilon_b}{d^3 \hat{\sigma}_{R,1}(\varepsilon)}} \phi(x_1)\, dx_1 \tag{47}$$

with the abbreviations

$$\hat{\sigma}_{R,1}(\varepsilon) = \left| \sigma_{R,1}(\varepsilon) - \sigma_{R,1}(\varepsilon_{\min}) \right| \tag{48}$$

and $x_1 = \xi_1 / d$. For the down-cycle in the same direction one finds a negative contribution to the total stress:

$$\sigma_{0,1}^{down}(\varepsilon) = \sigma_{R,1}(\varepsilon) - 2\tilde{\sigma}_{R,1}(\varepsilon) \int_{\frac{Q\varepsilon_b (1+\varepsilon_{\min})^{-3/2}}{2d^3 \tilde{\sigma}_{R,1}(\varepsilon_{\min})}}^{\frac{Q\varepsilon_b (1+\varepsilon)^{-3/2}}{2d^3 \tilde{\sigma}_{R,1}(\varepsilon)}} \phi(x_1)\, dx_1 \tag{49}$$

with the abbreviation

$$\tilde{\sigma}_{R,1}(\varepsilon) = \left| \sigma_{R,1}(\varepsilon) - \left(\frac{1+\varepsilon_{\max}}{1+\varepsilon} \right)^{3/2} \sigma_{R,1}(\varepsilon_{\max}) \right| \tag{50}$$

The negative sign in Eq. (49) results from the norm, implying, e.g., for the relative nominal stresses in direction 1 and 2:

$$\hat{\sigma}_{R,2}(\varepsilon) = 2(1+\varepsilon)^{3/2} \tilde{\sigma}_{R,1}(\varepsilon) \tag{51}$$

This is in contrast to the transformation behavior of the nominal stresses of the rubber matrix with a negative sign:

$$\sigma_{R,2}(\varepsilon) = -2(1+\varepsilon)^{3/2} \sigma_{R,1}(\varepsilon) \tag{52}$$

Note that the different choice of the extrema with $\partial\varepsilon_\mu/\partial t=0$ in Eqs. (48) and (50) are due to the fact that an up-cycle begins at $\varepsilon=\varepsilon_{min}$, but a down-cycle begins at $\varepsilon=\varepsilon_{max}$. As a rule, the relative stresses in the lower boundaries of the integrals in Eqs. (47) and (49) have to be chosen in such a way that they reach their maximum values, implying that all fragile clusters are broken and $\xi_\mu=\xi_{\mu,min}$. In deriving Eq. (49) we assumed an isotropic cluster size distribution, i.e., $\phi(x_1)=\phi(x_2)=\phi(x_3)$. This appears reasonable since $\phi(x_\mu)$ refers to the size distribution of clusters in the virgin, unstrained state. However, from the basic concept of cluster breakdown in the stretching direction and re-aggregation in the compression direction, respectively, it is clear that the size distribution of the filler clusters during the deformation cycles is strongly anisotropic. It leads to the characteristic hysteresis of a pre-strained sample, as given by Eqs. (47)–(50), even in the case of a quasi-static deformation at infinite slow strain rates.

Figure 45a–c shows an adaptation of the developed model to uniaxial stress-strain data of a pre-conditioned S-SBR-sample filled with 40 phr N220. The fits are obtained for the third stretching cycles at various pre-strains by referring to Eqs. (38), (44), and (47) with different but constant strain amplification factors $X=X_{max}$ for every pre-strain. For illustrating the fitting procedure, the adaptation is performed in three steps. Since the evaluation of the nominal stress contribution of the strained filler clusters by the integral in Eq. (47) requires the nominal stress $\sigma_{R,1}$ of the rubber matrix, this quantity is developed in the first step shown in Fig. 45a. It is obtained by demanding an intersection of the simulated curves according to Eqs. (38) and (44) with the measured ones at maximum strain of each strain cycle, where all fragile filler clusters are broken and hence the stress contribution of the strained filler clusters vanishes. The adapted polymer parameters are $G_c=0.176$ MPa and $n_e/T_e=100$, independent of pre-strain. According to the considerations at the end of Sect. 5.2.2, the tube constraint modulus is kept fixed at the value $G_e=0.2$ MPa, which is determined by the plateau modulus $G_N{}^o\approx0.4$ MPa [174, 175] of the uncross-linked S-SBR-melt ($G_e=1/2G_N{}^o$). The adapted amplification factors X_{max} for the different pre-strains ($\varepsilon_{max}=1$, 1.5, 2, 2.5, 3) are listed in the insert of Fig. 45a.

Figure 45b (upper part) shows the residual stress contribution of the strained filler clusters for the different pre-strains, obtained by subtracting the polymer contributions (solid lines) from the experimental stress-strain data (symbols) of Fig. 45a. The resulting data (symbols) are fitted to the second addend of Eq. (47) (solid lines), whereby the size distribution of filler clusters Eq. (37), shown in the lower part of Fig. 45b, has been used. The size distribution $\phi(x_1)$ is determined by the adapted mean cluster size $<x_1>\equiv<\xi_1/d>=26$ and the pre-chosen distribution width $\Omega=-0.5$, which allows for an analytical solution of the integral in Eq. (47). The tensile strength of filler-filler bonds is found as $Q\varepsilon_b/d^3=24$ MPa. The different fit lines result from the different stress-strain curves $\sigma_{R,1}(\varepsilon)$ that enter the upper boundary of the integral in Eq. (47). Note that this integral, representing the contribution of the strained filler clusters to the total stress, becomes zero at $\varepsilon=\varepsilon_{max}$ for every pre-strain.

Figure 45c shows the complete adaptation with Eq. (47) of the uniaxial stress-strain data, i.e., the sum of the two contributions shown in Fig. 45a,b. The fits for all pre-strained samples are very good in the large as well as the small strain regime, which is separately shown up to 150% strain in the insert of Fig. 45c. Furthermore, beside the third stretching cycles at various pre-strains, the first stretching of the virgin sample is shown. It is compared to a simulation curve given by Eq. (47) with the above specified material parameters and a strain dependent amplification factors $X=X(\varepsilon)$. Unfortunately, Eq. (39) cannot simply be applied for an estimation of $X(\varepsilon)$, since the upper boundary of the integral $\xi_\mu(\varepsilon)$ depends on $X(\varepsilon)$ itself (Eq. 34) and a general analytic solution of the integral is not known. For that reason an empirical power law dependency of the amplification factor X on a scalar strain variable E, involving the first deformation invariant $I_1(\varepsilon_\mu)$, has been applied [76, 89]:

$$X(E) = X_\infty + (X_o - X_\infty)(1 + E)^{-y} \tag{53}$$

with

$$E(\varepsilon_\mu) \equiv \left(\sum_{\mu=1}^{3} (1 + \varepsilon_\mu)^2 - 3 \right)^{1/2} \tag{54}$$

Here, X_∞ and X_o are the strain amplification factors at infinite and zero strain, respectively, and y is an empirical exponent. An adaptation of this empirical function, Eq. (53), to the X_{max}-values obtained for the pre-strained samples shown in Fig. 45a delivers the parameters X_o=11.5, X_∞=−1.21, and y=0.8, which are used for the simulation of the first stretching cycle shown in Fig. 45c.

A comparison of the experimental data for the first stretching cycle of the samples to the simulation curve in Fig. 45c shows no good agreement. Significant deviations are observed especially in the low strain regime, shown in the insert, where an extrapolation of the function $X(E)$ is used. The reason for the deviations may partly lie in the application of the power law approximation Eq. (53) for $X(E)$, instead of the micro-mechanically motivated Eq. (39). This may also lead to the unphysical negative value of X at extrapolated infinite strain (X_∞=−1.21).

A better fit between simulation and experimental data of the first extension of the virgin sample is obtained if the same model as above is used, but an empirical cluster size distribution is chosen instead of the physically motivated distribution function Eq. (37). This is demonstrated in Fig. 46a,b, where the adaptation of the same experimental data as above is made with a logarithmic normal form of the cluster size distribution function $\phi(x_1)$ with $x_1=\xi_1/d$:

$$\phi(x_1) = \frac{\exp\left(-\frac{\ln(x_1/<x_1>)^2}{2b^2}\right)}{\sqrt{\pi/2}\,b x_1} \tag{55}$$

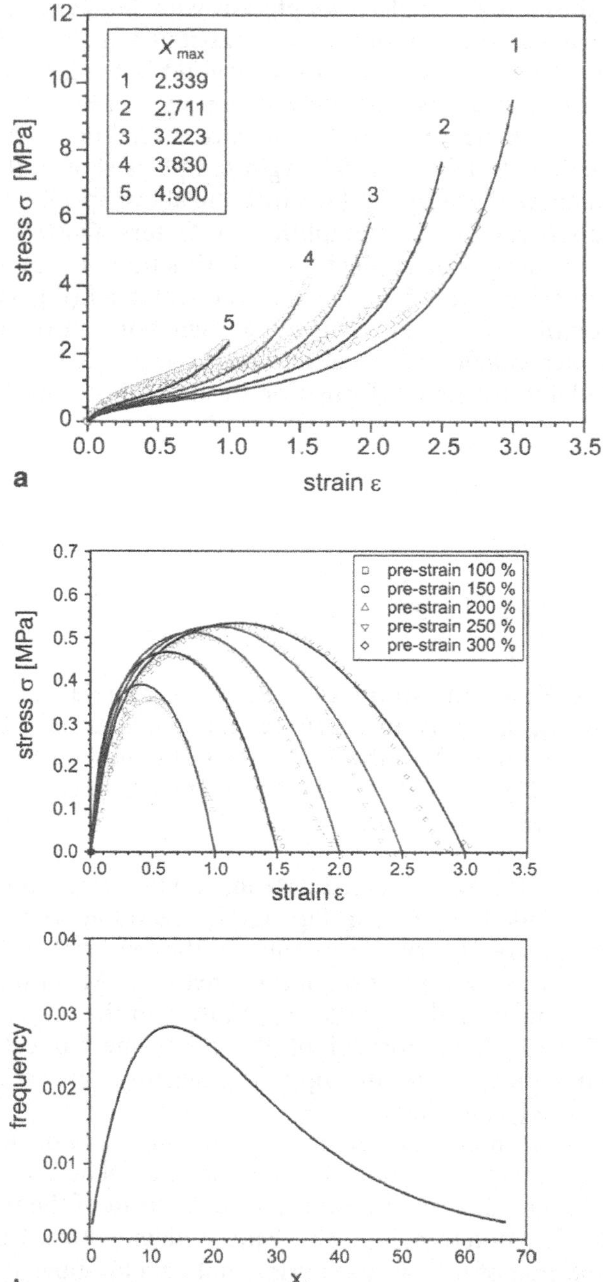

Fig. 45 **a** Uniaxial stress-strain data (*symbols*) of S-SBR samples filled with 40 phr N220 at various pre-strains ε_{max} and simulation curves (*lines*) of the polymer contribution according to Eqs. (38) and (44). The set of polymer parameters is found as G_c=0.176 MPa, G_e=0.2 MPa, and n_e/T_e=100. **b** Stress contributions of the strained filler clusters for the different pre-strains (*upper part*), obtained by subtracting the polymer contributions from the experimental stress-strain data of a. The *solid lines* are adaptations with the integral term of Eq. (47) and the cluster size distribution Eq. (37), shown

Figure 46a shows that a reasonable adaptation of the stress contribution of the clusters can be obtained, if the distribution function Eq. (55) with mean cluster size $<x_1>=25$ and distribution width $b=0.8$ is used. Obviously, the form of the distribution function is roughly the same as the one in Fig. 45b. The simulation curve of the first uniaxial stretching cycle now fits much better to the experimental data than in Fig. 45c. Furthermore, a fair simulation is also obtained for the equi-biaxial measurement data, implying that the "plausibility criterion", discussed at the end of Sect. 5.2.2, is also fulfilled for the present model of filler reinforced rubbers. Note that for the simulation of the equi-biaxial stress-strain curve, Eq. (47) is used together with Eqs. (45) and (38). The strain amplification factor $X(E)$ is evaluated by referring to Eqs. (53) and (54) with $\varepsilon_1=\varepsilon_2=\varepsilon$ and $\varepsilon_3=(1+\varepsilon)^{-2}-1$.

Similar well fitting simulation curves for the experimental stress-strain data as those shown in Fig. 46b can also be obtained for higher filler concentrations and silica instead of carbon black. In most cases, the log-normal distribution Eq. (55) gives a better prediction for the first stretching cycle of the virgin samples than the distribution function Eq. (37). Nevertheless, adaptations of stress-strain curves of the pre-strained samples are excellent for both types of cluster size distributions, similar to Fig. 45c and Fig. 46b. The obtained material parameters of four variously filled S-SBR composites used for testing the model are summarized in Table 4, whereby both cluster

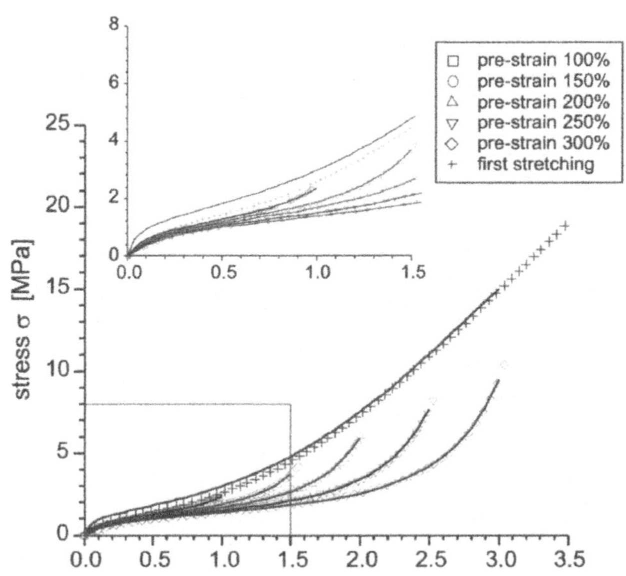

c

in the *lower part*. The obtained parameters of the filler clusters are $Q\varepsilon_b/d^3=24$ MPa, $<x_1>=26$, and $\Omega=-0.5$. c Uniaxial stress-strain data (*symbols*) of the S-SBR samples with 40 phr N220 at various pre-strains ε_{max} and for the first stretching, as indicated. The *insert* shows a magnification for the smaller strains. The *lines* are simulation curves with parameters specified in a,b

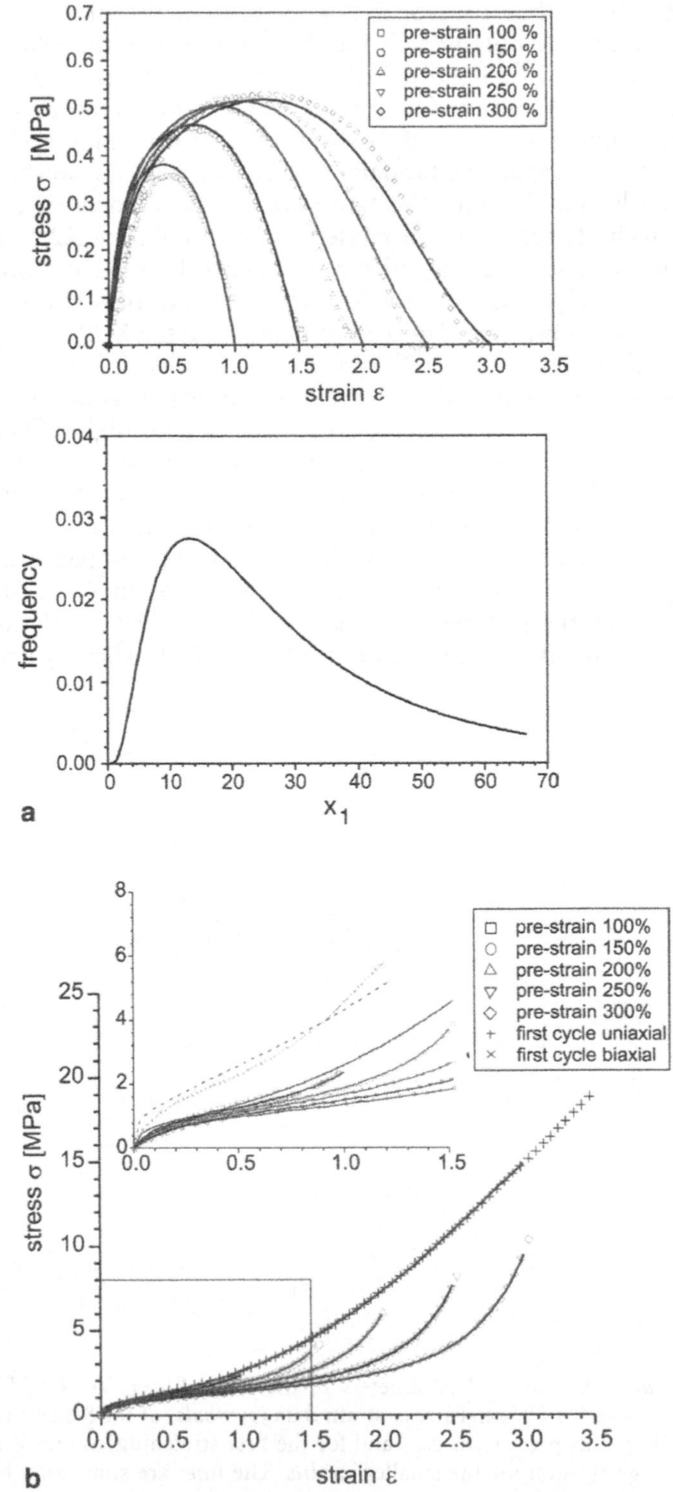

stress σ [MPa]
strain ε
pre-strain 100 %
pre-strain 150 %
pre-strain 200 %
pre-strain 250 %
pre-strain 300 %
frequency
x1
a
stress σ [MPa]
strain ε
pre-strain 100%
pre-strain 150%
pre-strain 200%
pre-strain 250%
pre-strain 300%
first cycle uniaxial
first cycle biaxial
b

Table 4 Material parameters obtained from adaptations of the model to uniaxial stress-strain data of four S-SBR samples filled with 40 and 60 phr N220 (C40, C60) and silica (S40, S60), respectively. Further polymer parameters are given as G_e=0.2 MPa and n_e/T_e=100, independent of sample type

		Cluster distribution Eq. (55)			Cluster distribution Eq. (37)			$X(E)$ Eq. (53)		
Sample type	G_c [MPa]	$<x>$	b	$Q\varepsilon_b/d^3$ [MPa]	$<x>$	Ω	$Q\varepsilon_b/d^3$ [MPa]	X_∞	X_o	y
C40	0.176	25	0.8	26	26	−0,5	24	−1.209	11.52	0.803
C60	0.190	18	0.8	26	20	−0,5	24	−0.329	17.40	1.180
S40	0.120	19	0.8	24	27	−0,5	28	−0.165	14.12	1.094
S60	0.130	22	0.8	24	31	−0,5	28	−0.148	15.24	1.130

size distributions, Eq. (37) and Eq. (55), have been considered. The obtained fit parameters for the polymer network, filler cluster morphology, and bond fracture mechanics appear physically suitable. In particular, the value of the cross-link modulus G_c increases with filler concentration due to the attachment of polymer chains at the filler surface, implying an increased effective cross-link density with rising filler loading. The G_c-values of the silica filled systems are found to be somewhat smaller than the carbon black filled systems, indicating that a higher fraction of curatives is deactivated by adsorption at the silica surface as compared to the carbon black surface. The estimated mean cluster size between $<\xi>\approx 20$ to 30 times the particle diameter d appears reasonable. With $d\approx 100$ nm one finds $<\xi>$ on the length scale of a few micrometers. Note that, due to Eq. (34), the quantity $Q\varepsilon_b/d^3$ represents the failure stress of filler-filler bonds, i.e., the limiting stress where all clusters are broken and the cluster size equals the particle size ($\xi=d$). The obtained fair agreement of the fitted values $Q\varepsilon_b/d^3\approx 24$ to 28 MPa with the tensile strength of the samples, which typically lies around 20 MPa, indicates that the failure stress of filler reinforced rubber samples is closely related to the failure stress of the filler-filler bonds. It demonstrates that the proposed mechanism of rubber reinforcement by stress-induced filler cluster breakdown can be considered to be responsible for the strongly enhanced tensile

Fig. 46 a Stress contributions of the strained filler clusters for the different pre-strains (*upper part*), obtained as in Fig. 45b. The *solid lines* are adapted with the integral term of Eq. (47) and the log-normal cluster size distribution Eq. (55), shown in the *lower part*. The obtained parameters of the filler clusters are $Q\varepsilon_b/d^3$=26 MPa, $<x_1>$=25, and b=0.8. **b** Uniaxial stress-strain data (symbols) as in Fig. 45c. The *insert* shows a magnification for the smaller strains, which also includes equi-biaxial data for the first stretching cycle. The *lines* are simulation curves with the log-normal cluster size distribution Eq. (55) and material parameters as specified in the *insert* of Fig. 45a and Table 4, sample type C40

strength of filled rubbers. Obviously, rupture of the whole sample appears somewhat before the last filler-filler bonds break down.

The success of the developed model in predicting uniaxial and equi-biaxial stress strain curves correctly emphasizes the role of filler networking in deriving a constitutive material law of reinforced rubbers that covers the deformation behavior up to large strains. Since different deformation modes can be described with a single set of material parameters, the model appears well suited for being implemented into a finite element (FE) code for simulations of three-dimensional, complex deformations of elastomer materials in the quasi-static limit.

Beside the consideration of the up-cycles in the stretching direction, the model can also describe the down-cycles in the backwards direction. This is depicted in Fig. 47a,b for the case of the S-SBR sample filled with 60 phr N 220. Figure 47a shows an adaptation of the stress-strain curves in the stretching direction with the log-normal cluster size distribution Eq. (55). The depicted down-cycles are simulations obtained by Eq. (49) with the fit parameters from the up-cycles. The difference between up- and down-cycles quantifies the dissipated energy per cycle due to the cyclic breakdown and re-aggregation of filler clusters. The obtained microscopic material parameters for the viscoelastic response of the samples in the quasi-static limit are summarized in Table 4.

A comparison of the simulation curves in Fig. 47a with experimental stress-strain cycles, depicted, e.g., in Fig. 3, shows a significantly different behavior of the down-cycles. Obviously, the evaluated stress contribution of the strained filler clusters during the down cycle, i.e., the integral term of Eq. (49), is too small to explain the difference between the experimentally observed up- and down-cycles in the high strain regime, where the upturn of the stress appears. The reason for this deviation can partly be found in the pronounced set behavior of reinforced rubbers that shifts the remaining strain at zero stress to increasing positive values with increasing pre-strain (compare Fig. 3). So far, this set behavior is not considered in the developed model. Concerning the measurement results, it has been compensated by shifting the experimental stress-strain curves of the pre-strained samples into the origin. This procedure is described more closely in [76].

In view of an illustration of the viscoelastic characteristics of the developed model, simulations of uniaxial stress-strain cycles in the small strain regime have been performed for various pre-strains, as depicted in Fig. 47b. Thereby, the material parameters obtained from the adaptation in Fig. 47a (Table 4, sample type C60) have been used. The dashed lines represent the polymer contributions, which include the pre-strain dependent hydrodynamic amplification of the polymer matrix. It becomes clear that in the small and medium strain regime a pronounced filler-induced hysteresis is predicted, due to the cyclic breakdown and re-aggregation of filler clusters. It can considered to be the main mechanism of energy dissipation of filler reinforced rubbers that appears even in the quasi-static limit. In addition, stress softening is present, also at small strains. It leads to the characteristic decline of the polymer contributions with rising pre-strain (dashed lines in

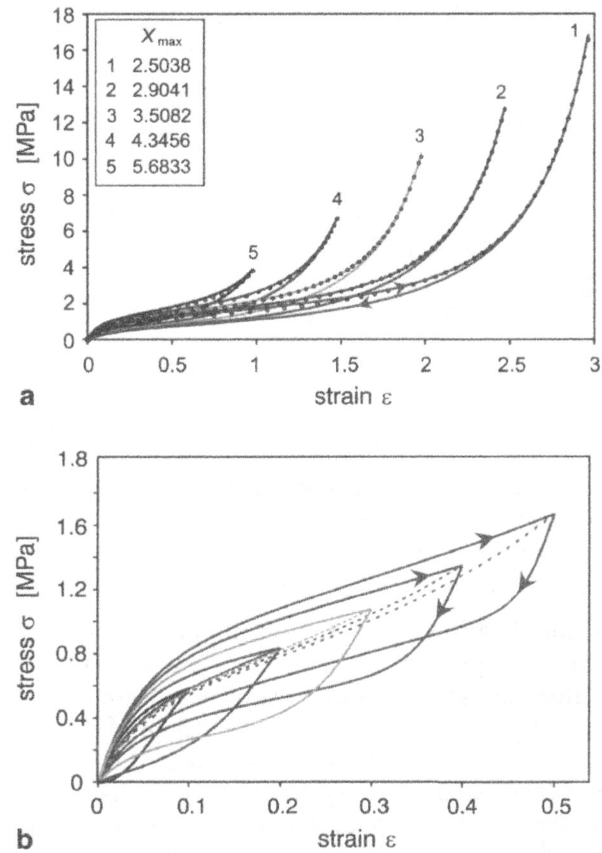

Fig. 47 **a** Uniaxial stress-strain data in stretching direction (*symbols*) of S-SBR samples filled with 60 phr N 220 at various pre-strains ε_{max} and simulations (*solid lines*) of the third up- and down-cycles with the cluster size distribution Eq. (55). Fit parameters are listed in the *insert* and Table 4, sample type C60. **b** Simulation of uniaxial stress-strain cycles for various pre-strains between 10 and 50% (*solid lines*) with material parameters from the adaptation in a. The *dashed lines* represent the polymer contributions according to Eqs. (38) and (44) with different strain amplification factors

Fig. 47b), which impacts the slope of the stress-strain cycles. This softening effect results from the drop of the strain amplification factor X_{max} with increasing pre-strain, which has been determined by an extrapolation of the adapted values, shown in the insert of Fig. 47a, with the power law approximation Eq. (53).

It is important to note that stress softening is also present during dynamic stress-strain cycles of filled rubbers at small and medium strain. In particular, this can be concluded from the dynamic mechanical data of the S-SBR samples filled with 60 phr N 220 as shown in Fig. 48. In the framework of the above model, the observed shift of the center point of the cycles to smaller stress values with increasing strain amplitude or maximum strain and the accompanied drop of the slope of the hysteresis cycles can be related to a de-

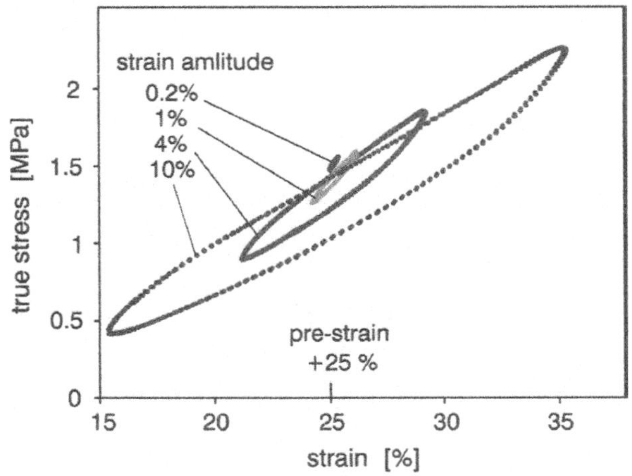

Fig. 48 Dynamic stress-strain cycles obtained from uniaxial harmonic excitations at 1 Hz of the S-SBR samples filled with 60 phr N 220 at fixed pre-strain and various strain amplitudes, as indicated

creasing strain amplification factor X_{max}. It implies that a pronounced Payne effect appears also for pre-strained samples that compares to the one observed for dynamic excitations around the stress-strain origin. This has led to some conceptual problems in relating the Payne effect to filler network breakdown [176, 177], which is briefly reviewed in the introduction section of [57]. The problems arise primary from the fact that the shifting of the center points of the hysteresis cycles due to stress softening is ignored if the concept of linear viscoelasticity is applied and only the dynamic storage and loss moduli are considered.

The experimental stress-strain cycles shown in Fig. 48 are quite similar to the simulation curves in Fig. 47b, though the stress level of the experimental curves is somewhat larger. This can partly be related to the fact that the true stress is shown, which is by a factor of about 1.25 larger than the apparent stress. Nevertheless, it appears that the extrapolated strain amplification factors X_{max}, used for the simulation, are too small and a separate adaptation for the small strain regime is needed. However, at present an adaptation of the developed model to the dynamic data is not possible, mainly for two reasons. On the one hand, an explicit time dependency of the stress-strain cycles is not considered so far. On the other, only full cycles from maximum to minimum strain can be described, where all fragile clusters break down and re-combine. A more general model must also consider the re-aggregation mechanism in dependence of strain, in order to describe all kind of stress-strain cycles. This will be a task of future work. Despite these limitations there is strong evidence that the Payne effect can be described on the basis of the derived model of stress softening and filler-induced hysteresis, which allows for a proper simulation of quasi-static stress-strain curves, referred to as Mullins effect. This demonstrates that both effects can be traced back to the same micro-mechanical mechanism.

6 Summary and Conclusions

In the first part of the chapter (Sect. 3), the morphology and surface activity of carbon black, the most important filler in the rubber industry, are analyzed. It is shown that, due to the disordered nature of carbon black formation during processing in a furnace reactor, the morphology of furnace blacks is well described by referring to a fractal analysis. Investigations of the surface roughness of carbon blacks by static gas adsorption point out that all furnace blacks exhibit a pronounced surface roughness with an almost unique value of the surface fractal dimension $d_s \approx 2.5$–2.6 on atomic length scales below 6 nm. This universality is related to a particular random deposition mechanism of carbon nuclei on condensed carbon particles that governs the surface growth of carbon blacks during processing. Graphitization of carbon blacks diminishes the roughness of the surface on length scales below $z<1$ nm, while on larger length scales the surface fractal dimension remains unchanged. The surface fractal dimension of all examined graphitized blacks is found as $d_s \approx 2.3$ below 1 nm, independent of the applied evaluation procedure.

An analysis of the surface energy distribution of furnace blacks demonstrates that four different energetic sites can be distinguished. They are shown to be related to the characteristic surface morphology of carbon blacks. The fraction of highly energetic sites decreases significantly with grade number and disappears almost completely during graphitization. It indicates that the reinforcing potential of carbon black is closely related to the amount of highly energetic sites that can be well quantified by the applied gas adsorption technique at low pressure. According to the universal surface roughness of furnace blacks independent of grade number, one expects that the reinforcing potential owing to the surface roughness of carbon blacks is not much different. Nevertheless, the detected high energetic and morphological disorder of the carbon black surface can considered to be responsible for a strong polymer-filler coupling and phase bonding. Beside the specific surface, this is an important factor for obtaining adequate ultimate properties of carbon black reinforced rubbers.

The morphology of the more or less ramified primary carbon black aggregates, as analyzed by TEM techniques, is readily described by a fractal approach. This refers to the specific cluster-cluster aggregation mechanism of primary particles under ballistic conditions, which is realized during carbon black processing. It implies a mass fractal dimension $d_f \approx 1.9$–1.95, which is indeed found for the fine blacks. For the more coarse furnace blacks larger values of d_f are obtained that increase with the grade number. This results from electrostatic repulsive interactions between the aggregates due to the application of processing agents. During compounding of carbon blacks with highly viscous rubbers, rupture of primary aggregates takes place depending on the mixing severity and grade number. It implies that the aggregate size decreases with increasing mixing time or filler loading, preferably for the high structured coarse blacks. The resulting influence on the mechanical properties of the composites can be described on a quantitative level by considering the rising solid fraction of primary aggregates with increasing mixing severity, which impacts the mechanically effective filler concentration.

In the second part of the chapter (Sect. 4), carbon black networking in elastomer composites is analyzed by applying TEM-, electrical percolation-, dielectric- as well as flocculation investigations. This provides information on the fractal nature of filler networks as well as the morphology of filler-filler bonds. It becomes obvious that TEM analysis gives a limited microscopic picture of the filler network morphology, only, due to spatial interpenetration of flocculated neighboring filler clusters. Investigations of the d.c.-conductivity demonstrate that no universal percolation structure for the filler network is realized on mesoscopic length scales if a critical filler concentration is exceeded. Instead, the structure appears to be modified by a superimposed kinetic aggregation process, which explains the detected non-universal value of the percolation exponent and the impact of filler specific surface and polymer micro-structure on the percolation threshold. Flocculation studies, considering the small strain mechanical response of the uncross-linked composites during heat treatment (annealing), demonstrate that a relative movement of the particles takes place that depends on particle size, molar mass of the polymer, as well as polymer-filler and filler-filler interaction. This provides strong experimental evidence for a kinetic cluster-cluster aggregation mechanism of filler particles in the rubber matrix to form a filler network. The a.c.-conductivity exponent in the high frequency regime is shown to be related to an anomalous diffusion mechanism of charge carriers on fractal carbon black clusters. This confirms the fractal nature of filler networks in elastomers below a certain length scale, though it gives no definite information on the particular network structure.

From the dielectric investigations it becomes obvious that charge transport above the percolation threshold is limited by a hopping or tunneling mechanism of charge carriers over small gaps of order 1 nm between adjacent carbon black particles. From this finding and the observed dependency of the flocculation dynamics on the amount of bound rubber, a model of filler-filler bonds is developed. It relates the mechanical stiffness of filler-filler bonds to the remaining gap size between the filler particles that develops during annealing (and cross-linking) of filled rubbers. In this model, stress between adjacent filler particles in a filler cluster is assumed to be transmitted by nanoscopic, flexible bridges of glassy-like polymer, implying that a high flexibility and strength of filler clusters in elastomers is reached. This picture of filler-filler bonds allows for a qualitative explanation of the observed flocculation effects by referring to the amount of bound rubber and it is impact on the stiffness and strength of filler-filler bonds.

In the last part of the chapter (Sect. 5), a micro-mechanical model of rubber reinforcement by flexible filler clusters is developed that allows for a quantitative analysis of stress softening and hysteresis of filler reinforced rubbers up to large quasi-static deformations. It is shown that the kinetic cluster-cluster aggregation (CCA) model of filler networking in elastomers represents a reasonable theoretical basis for understanding the small strain viscoelastic properties of reinforced rubbers. According to this model, filler networks consist of a space-filling configuration of CCA-clusters with characteristic mass fractal dimension $d_f \approx 1.8$. The consideration of flexible chains of filler particles, approximating the elastically effective backbone of the filler clusters, allows for a micro-mechanical description of the elastic properties of tender CCA-clusters in elastomers. The main contribution of the elastically stored energy in the strained filler clusters results from the bending-twisting deformation of filler-filler bonds. The predicted power-law behavior of the small strain modulus of filler reinforced rubbers is confirmed by a variety of experimental data, including carbon black and silica filled rubbers as well as composites with model fillers (microgels).

Beside the elastic behavior, the failure properties of filler-filler bonds and filler clusters are considered in dependence of cluster size. This completes the micro-mechanical description of tender but fragile filler clusters in the stress field of a strained rubber matrix. By assuming a specific cluster size distribution in reinforced rubbers, which is motivated by an asymptotic solution of Smoluchowski's equation for the kinetics of irreversible cluster-cluster aggregation, a constitutive material model of filler reinforced rubbers is derived. It is based on a non-affine tube model of rubber elasticity, including hydrodynamic amplification of the rubber matrix by a fraction of rigid filler clusters with filler-filler bonds in the unbroken, virgin state. The filler-induced hysteresis is described by an anisotropic free energy density, considering the cyclic breakdown and re-aggregation of the residual fraction of more fragile filler clusters with already broken filler-filler bonds. The model assumes that the breakdown of filler clusters during the first deformation of the virgin samples is totally reversible, though the initial virgin state of filler-filler bonds is not recovered. This implies that, on the one hand, the fraction of rigid filler clusters decreases with increasing pre-strain, leading to the pronounced stress softening after the first deformation cycle. On the other hand, the fraction of fragile filler clusters increases with increasing pre-strain, which impacts the filler-induced hysteresis.

It is demonstrated that the quasi-static stress-strain cycles of carbon black as well as silica filled rubbers can be well described in the scope of the theoretic model of stress softening and filler-induced hysteresis up to large strain. The obtained microscopic material parameter appear reasonable, providing information on the mean size and distribution width of filler clusters, the tensile strength of filler-filler bonds, and the polymer network chain density. In particular it is shown that the model fulfils a "plausibility criterion" important for FE applications. Accordingly, any deformation mode can be predicted based solely on uniaxial stress-strain measurements, which can be carried out relatively easily.

From the simulations of stress-strain cycles at small and medium strain it can be concluded that the model of anisotropic cluster breakdown and re-aggregation for pre-strained samples represents a fundamental micro-mechanical basis for the description of non-linear viscoelasticity of filler reinforced rubbers. Thereby, the mechanisms of energy storage and dissipation are traced back to the elastic response of the polymer network as well as the elasticity and fracture properties of flexible filler clusters. For a time dependent, complete characterization of non-linear viscoelastic stress-strain cycles of filler reinforced rubbers, the dynamic-mechanical response of the polymer matrix has to be considered as well. Furthermore, the re-aggregation rate must be specified in dependence of strain, in order to describe any kind of stress-strain cycles. This will be a task of future work.

Acknowledgements It is a great pleasure to thank Dr. G. Heinrich (Continental AG) and Prof. R. H. Schuster (DIK) for continuous support and many inspiring discussions. I'm also grateful to Dr. M. Gerspacher (Sid Richardson Carbon Co.) for the very fruitful cooperation and for providing electrical data. Helpful discussions with Prof. Th. Vilgis (MPI Mainz), Prof. E. Straube (University Halle-Wittenberg), and Prof. W. Gronski (University Freiburg) are appreciated.

Special thanks are extended to Dr. A. Schröder (Freudenberg AG) and Dr. J. Schramm (Continental AG) for the fruitful cooperation during their employment at the DIK. The cooperation and friendly help of Dr. J. Meier, Dr. Th. Alshuth, Dr. H. Geisler, and Dipl. Phys. A. Müller (all DIK) is highly appreciated.

Support by the Deutsche Kautschukgesellschaft e. V., the Luise Arntz Stiftung, Continental AG, Bayer AG, Columbian Chemicals Co., Sid Richardsen Carbon Co., and Degussa AG is gratefully acknowledged.

References

1. Mandelbrot BB (1977) Fractals: form, chance and dimension. WH Freeman, San Francisco
2. Mandelbrot BB (1982) The fractal geometry of nature. WH Freeman, New York
3. Meakin P (1990) Prog Solid State Chem 20:135
4. Lin MY, Lindsay HN, Weitz DA, Ball RC, Klein R (1989) Nature 339:360
5. Witten TA, Cates ME (1986) Science 232:1607
6. Bunde A, Havlin S (eds) (1991) Fractals and disordered systems. Springer, Berlin Heidelberg New York
7. Bunde A, Havlin S (eds) (1994) Fractals in science. Springer, Berlin Heidelberg New York
8. Jullien B (1990) New J Chem 14:239
9. Witten TA, Sander LM (1981) Phys Rev Lett 47:1400; Phys Rev B Condens Matter B27:5686
10. Meakin P (1984) Phys Rev A 29:997
11. Ball RC, Jullien R (1984) J Phys (Paris) Lett 45:L103
12. Meakin P (1988) Adv Colloid Interface Sci 28:249
13. Weitz DA, Oliveria M (1983) Phys Rev Lett 52:1433
14. Meakin P, Donn B, Mulholland GW (1989) Langmuir 5:510
15. Megaridis CM, Dobbins RA (1990) Combust Sci Technol 71:95
16. Samson RJ, Mulholland GW, Gentry JW (1987) Langmuir 3:272
17. Sutherland DN (1970) Nature 226:1241
18. Heinrich G, Klüppel M (2001) Kautschuk Gummi Kunstst 54:159
19. Barabasi AL, Stanley HE (1996) Fractal concepts in surface growth. Cambridge University Press; Barabasi AL, Araujo M, Stanley HE (1992) Phys Rev Lett 68:3729

20. Schröder A, Klüppel M, Schuster RH, Heidberg J (2001) Kautschuk Gummi Kunstst 54:260
21. Witten TA, Rubinstein M, Colby RH (1993) J Phys II (France) 3:367
22. Klüppel M, Heinrich G (1995) Rubber Chem Technol 68:623
23. Klüppel M, Schuster RH, Heinrich G (1997) Rubber Chem Technol 70:243
24. Heinrich G, Straube E, Helmis G (1988) Adv Polym Sci 85:33
25. Brown D (1987) PhD thesis, University of Cambridge, UK
26. Donnet JB, Bansal RC, Wang MJ (eds) (1993) Carbon black: science and technology. Marcel Decker, New York Hongkong
27. Kraus G (ed) (1965) Reinforcement of elastomers. Interscience Publ, New York London Sydney
28. Payne AR (1962) J Appl Polym Sci 6:57
29. Payne AR (1963) J Appl Polym Sci 7:873
30. Payne AR (1964) J Appl Polym Sci 8:2661
31. Payne AR (1964) Trans IRI 40:T135
32. Payne AR (1965) In: Kraus G (ed) Reinforcement of elastomers. Interscience Publ, New York London Sydney
33. Payne AR (1965) J Appl Polym Sci 9:2273,3245
34. Payne AR (1972) J Appl Polym Sci 16:1191
35. Payne AR (1963) Rubber Chem Technol 36:432
36. Medalia AI (1973) Rubber Chem Technol 46:877
37. Medalia AI (1974) Rubber Chem Technol 47:411
38. Voet A, Cook FR (1967) Rubber Chem Technol 40:1364
39. Voet A, Cook FR (1968) Rubber Chem Technol 41:1215
40. Dutta NK, Tripathy DK (1989) Kautschuk Gummi Kunstst 42:665
41. Dutta NK, Tripathy DK (1992) J Appl Polym Sci 44:1635
42. Dutta NK, Tripathy DK (1990) Polym Test 9:3
43. Ulmer JD, Hergenrother WL, Lawson DF (1998) Rubber Chem Technol 71:637
44. Wang MJ, Patterson WJ, Ouyang GB (1998) Kautschuk Gummi Kunstst 51:106
45. Freund B, Niedermeier W (1998) Kautschuk Gummi Kunstst 51:444
46. Mukhopadhyay K, Tripathy DK (1992) J Elastomers Plast 24:203
47. Wang MJ (1998) Rubber Chem Technol 71:520
48. Bischoff A, Klüppel M, Schuster RH (1998) Polym Bull 40:283
49. Vieweg S, Unger R, Heinrich G, Donth E (1999) J Appl Polym Sci 73:495
50. Payne AR, Watson WF (1963) Rubber Chem Technol 36:147
51. Amari T, Mesugi K, Suzuki H (1997) Prog Org Coat 31:171
52. Payne AR, Wittaker RE (1970) Rheol Acta 9:91
53. Payne AR, Wittaker RE (1970) Rheol Acta 9:97
54. Kraus G (1984) J Appl Polym Sci Appl Polym Symp 39:75
55. van de Walle A, Tricot C, Gerspacher M (1996) Kautsch Gummi Kunstst 49:173
56. Lin CR, Lee YD (1996) Macromol Theory Simul 5:1075; (1997) Macromol Theory Simul 6:102
57. Heinrich G, Klüppel M (2002) Adv Polym Sci 160:1
58. Mullins L (1948) Rubber Chem Technol 21:281
59. Mullins L, Tobin RN (1957) Rubber Chem Technol 30:355
60. Mullins L (1965) In: Kraus G (ed) Reinforcement of elastomers,. Interscience Publ, New York London Sydney
61. Medalia AI (1978) Rubber Chem Technol 51:437
62. Edwards SF, Vilgis TA (1988) Rep Prog Phys 51:243; (1986) Polymer 27:483
63. Guth E, Gold O (1938) Phys Rev 53:322
64. Huber G (1997) PhD thesis, University of Mainz, Germany
65. Huber G, Vilgis TA (1998) Euro Phys J B3:217; (1999) Kautschuk Gummi Kunstst 52:102
66. Heinrich G, Vilgis TA (1993) Macromolecules 26:1109
67. Eisele U, Müller HK (1990) Kautschuk Gummi Kunstst 43:9
68. Bueche F (1960) J Appl Polym Sci 4:107; (1961) J Appl Polym Sci 5:271

69. Govindjee S, Simo J (1991) J Mech Phys Solids 39:87; (1992) J Mech Phys Solids 40:213
70. Dannenberg EM (1974) Rubber Chem Technol 47:410
71. Rigbi Z (1980) Adv Polym Sci 36:21
72. Hamed GR, Hatfield S (1989) Rubber Chem Technol 62:143
73. Haarwood JA, Mullins L, Payne AR (1965) J Appl Polym Sci 9:3011
74. Haarwood JA, Payne AR (1966) J Appl Polym Sci 10:315, 1203
75. Klüppel M, Schramm J (1999) In: Dorfmann A, Muhr A (eds.) Constitutive models for rubber. Balkema, Rotterdam, p 211
76. Klüppel M, Schramm J (2000) Macromol Theory Simul 9:742
77. Kaliske M, Heinrich G (1999) Rubber Chem Technol 72:602
78. Heinrich G, Kaliske M (1997) Comput Theor Polym Sci 7:227
79. Schröder A (2000) PhD thesis, University of Hannover, Germany
80. Schröder A, Klüppel M, Schuster RH (1999) Kautschuk Gummi Kunstst 52:814
81. Schröder A, Klüppel M, Schuster RH (2000) Kautschuk Gummi Kunstst 53:257
82. Schröder A, Klüppel M, Schuster RH, Heidberg J (2002) Carbon 40:207
83. Klüppel M, Schröder A, Schuster RH, Schramm J (2000) Paper No XLI, ACS Rubber Division Meeting, Dallas, USA
84. Klüppel M, Schröder A, Heinrich G, Schuster RH, Heidberg J (2000) Proceedings: Kautschuk-Herbst-Kolloquium, Hannover, Germany, p 193
85. Schuster RH, Klüppel M, Schramm J, Heinrich G (1998) Paper No 56, ACS Rubber Division Meeting, Indianapolis, USA
86. Klüppel M, Schuster RH, Schaper J (1999) Rubber Chem Technol 72:91
87. Schuster RH, Meier J, Klüppel M (2000) Kautschuk Gummi Kunstst 53:663
88. Schuster RH, Klüppel M, Früh T (1997) Paper No 53, ACS Rubber Division Meeting, Anaheim, USA
89. Klüppel M, Meier J (2001) In: Besdo D, Schuster RH, Ihlemann J (eds.) Constitutive models for rubber II. Balkema, Lisse Abingdon Exton Tokyo, p 11
90. Raab H, Fröhlich J, Göritz D (2000) Kautschuk Gummi Kunstst 53:137
91. Niedermeier W, Stierstorfer J, Kreitmeier S, Metz O, Göritz D (1994) Rubber Chem Technol 67:148
92. Niedermeier W, Raab H, Stierstorfer J, Kreitmeier S, Göritz D (1994) Kautschuk Gummi Kunstst 47:799
93. Hjelm RP, Wampler WA, Serger PA (1991) Paper No 5, ACS Rubber Division Meeting, Toronto, Canada
94. Rieker TP, Misono S, Ehrenbourger-Dolle F (1999) Langmuir 15:914
95. Rieker TP, Hindermann-Bischoff M, Ehrenbourger-Dolle F (2000) Landmuir 16:5588
96. Fröhlich J, Kreitmeier S, Göritz D (1998) Kautschuk Gummi Kunstst 51:370
97. Zerda TW, Yang H, Gerspacher M (1992) Rubber Chem Technol 65:130
98. Isamail JMK, Pfeifer P (1994) Langmuir 10:1532
99. Isamail JMK (1992) Langmuir 8:360
100. Pfeifer P, Obert M, Cole MW (1989) Proc R Soc Lond A423:169
101. Pfeifer P, Cole MW (1990) New J Chem 14:221
102. Lippens BC, Linsen BG, De Boer JH (1964) J Catalysis 3:32
103. Adamson AW, Ling L (1961) Adv Chem 33:51
104. Langmuir I (1918) J Am Chem Soc 40:1316.
105. Fowler RH, Guggenheim EA (1952) Statistical thermodynamics. Cambridge University Press, Cambridge, UK
106. Douglas JF (1989) Macromolecules 22:3707
107. Vilgis TA, Heinrich G (1994) Macromolecules 27:7846
108. Herd CR, McDonald GC, Hess WM (1992) Rubber Chem Technol 65:107
109. Gerspacher M, O'Farrell CP (1992) Kautschuk Gummi Kunstst 45:97
110. Le Mehaute A, Gerspacher M, Tricot C (1993) In: Donnet JB, Bansal RC, Wang MJ (eds) Carbon black: science and technology. Marcel Decker, New York Hongkong, p 245
111. Medalia AI (1967) J Colloid Interface Sci 24:393

112. Medalia AI, Heckman FA (1996) Carbon 7:567
113. Viswanathan R, Heaney MB (1995) Phys Rev Lett 75:4433
114. Ehrburger-Dolle F, Tence M (1990) Carbon 28:448
115. Oppermann B (1994) PhD thesis, University of Hannover, Germany
116. Heckman FA, Medalia AI (1969) J Inst Rubber Ind 3:66
117. Hess WM, Chirico VE, Burgess KA (1973) Proceedings: International Rubber Conference, Prague, CR
118. Gessler M (1970) Rubber Chem Technol 43:943
119. Sgodzaj U (2001) PhD thesis, University of Karlsruhe, Germany
120. Jäger KM, McQueen DH (2001) Polymer 42:9575
121. Medalia AI (1986) Rubber Chem Technol 59:432
122. Boettger H, Bryksin UV (1986) Hopping conduction in solids. Verlag Akademie, Berlin
123. Miyasaka K (1985) Nippon Gomu Kyokaishi 58:561
124. O'Farrell CP, Gerspacher M, Nikiel L (2000) Kautschuk Gummi Kunstst 53:701
125. Probst N, Grivei E (2000) Proceedings: Kautschuk-Herbst-Kolloquium, Hannover, Germany, p 205
126. Balberg I (1998) Trends Stat Phys 2:39; (1987) Phys Rev Lett 59:1305
127. Rubin Z, Sunshine A, Heaney MB, Bloom I, Balberg I (1999) Phys Rev B59:12,196
128. Stauffer D, Aharoni A (1992) Introduction to percolation theory. Taylor and Francis, London
129. Ayala JA, Hess WM, Dotson AO, Joyce GA (1990) Rubber Chem Technol 63:747
130. Hess WM, Chirico VE (1975) Proceedings: Plast Rubber Inst 1st Eur Conf, Brussels, Belgum
131. van Beek LKH (1967) In: Birks JB (ed.) Progress in dielectrics. vol 7. Heywood, London, p. 69
132. Ouyang GB (2000) Proceedings: Kautschuk-Herbst-Kolloquium, Hannover, Germany, p 183
133. Schwartz G, Cerveny S, Marzocca AJ (2000) Polymer 41:6589
134. Kremer F, Ezquerra TA, Mohamadi M, Bauhofer W, Vilgis TA, Wegner G (1988) Solid State Commun 66:153
135. Jäger KM, McQueen DH, Tchmutin IA, Ryvtina NG, Klüppel M (2001) J Phys D Appl Phys 34:2699
136. Gefen Y, Aharony A, Alexander S (1983) Phys Rev Lett 50:77
137. Havlin S, Bunde A (1991) In: Bunde A, Havlin S (eds) Fractals and disordered systems. Springer, Berlin Heidelberg New York, p 115
138. Kastner A, Alig I, Heinrich G, Klüppel M (In preparation)
139. Klüppel M (1997) Kautschuk Gummi Kunstst 50:282
140. Georg G, Böhm A, Nguyen M (1995) J Appl Polym Sci 55:1041
141. Fröhlich J, Niedermeier W (2000) Proceedings: German Rubber Conference, Nürnberg, Germany, p 107
142. Fröhlich J, Luginsland HD (2000) Proceedings: Kautschuk-Herbst-Kolloquium, Hannover, Germany, p 13
143. Wang T, Wang MJ, Shell J, Tokita N (2000) Kautschuk Gummi Kunstst 53:497
144. Schuster RH (1989) Verstärkung von Elastomeren durch Ruß, WDK-Grünes Buch Nr 40
145. Cotten GR (1975) Rubber Chem Technol 48:548
146. Kraus G, Gruver JT (1968) Rubber Chem Technol 41:1256
147. Meissner B (1974) J Appl Polym Sci 18:2483
148. Brennan JJ, Jermyn TE, Boonstra BB (1965) Gummi Asbest Kunstst 18:266
149. Boonstra BB (1967) J Appl Polym Sci 11:389
150. Sircar AK, Voet A (1970) Rubber Chem Technol 43:973
151. Wolff S, Wang MJ, Tan EH (1993) Rubber Chem Technol 66:163
152. Chapman AV, Fulton WS, Tinker AJ (2000) Proceedings: Kautschuk-Herbst-Kolloquium, Hannover, Germany, p 23
153. Kida N, Ito M, Yatsunyanagi F, Kaido H (1996) J Appl Polym Sci 61:1345

154. Sircar AK, Lamond TG (1975) Rubber Chem Technol 48:79, 89
155. Kantor Y, Webman I (1984) Phys Rev Lett 52:1891
156. Friedlander SK, Jang HD, Ryu KH (1998) Appl Phys Lett 72:173
157. Friedlander SK, Ogawa K, Ullmann M (2001) J Polym Sci Polym Phys 38:2658
158. Müller M (2002) PhD thesis, University of Hannover, Germany
159. Müller M, Schuster RH (2000) Proceedings: German Rubber Conference, Nürnberg, Germany, p 77
160. Tschoegl NW (1989) The phenomenological theory of linear viscoelastic behavior. Springer, Berlin Heidelberg New York
161. van Dongen PGJ, Ernst MH (1985) Phys Rev Lett 54:1396; (1985) J Phys A18:2779
162. Ziff RM, McGrady ED, Meakin P (1985) J Chem Phys 82:5269
163. Heinrich G, Straube E (1984) Acta Polym 35:115
164. Heinrich G, Straube E (1987) Polym Bull 17:247
165. Rubinstein M, Panyukow S (1997) Macromolecules 30:8036
166. Straube E, Urban V, Pyckhout-Hintzen W, Richter D, Glinka CW (1995) Phys Rev Lett 74:4464
167. Klüppel M (1994) Macromolecules 27:7179
168. Klüppel M, Menge H, Schmidt H, Schneider H, Schuster RH (2001) Macromolecules 34:8107
169. Edwards SF, Vilgis TA (1988) Rep Prog Phys 51:243; (1986) Polymer 27:483
170. Klüppel M, Heinrich G (1994) Macromolecules 27:3569
171. Fetters LJ, Lohse DJ, Richter D, Witten TA, Zirkel A (1994) Macromolecules 27:4639
172. Klüppel M (1992) Prog Colloid Polym Sci 90:137
173. Klüppel M (1993) J Appl Polym Sci 48:1137
174. Meier J (2002) PhD thesis, Technical University of Clausthal, Germany
175. Meier J, Giebeler E, Klüppel M, Schuster RH (2001) Paper No. XX, ACS Rubber Division Meeting, Provicence, USA
176. Brown JD (1997) PhD thesis, Rensselaer Polytechnic Institute, Troy, New York, USA
177. Chazeau L, Brown JD, Yanyo LC, Sternstein SS (2000) Polym Compos 21:202

Editor: Karel Dušek

Received: August 2002

Adv Polym Sci (2003) 164:87–210
DOI 10.1007/978-3-540-36438-2

Materials Contrasts and Nanolithography Techniques in Scanning Force Microscopy (SFM) and their Application to Polymers and Polymer Composites

Martin Munz · Brunero Cappella · Heinz Sturm · Markus Geuss · Eckhard Schulz

Federal Institute for Materials Research and Testing (BAM), Unter den Eichen 87, 12205 Berlin, Germany
E-mail: heinz.sturm@bam.de

Abstract Beyond measuring the topography of surfaces, scanning force microscopy (SFM) has proved to be valuable both for mapping of various materials properties and for modifying surfaces via lithography techniques. Thus, SFM has gained relevance as a surface analysis technique as well as a tool for nanoscale engineering purposes. Different kinds of tip-sample interactions are exploitable, e.g. mechanical, thermal and electrical ones. Owing to its versatility, SFM has found plenty of applications in polymer science. Among others, the examples reported on in this review article encompass issues related to commodity polymers, various polymer-based composites, polymer blends, or ferroelectric polymers. For instance, stiffness imaging is elucidated as a technique for detecting interphases occurring in composites with inorganic fillers. The described applications are mainly related to mechanical and electrical tip-sample interactions. The respective fundamentals are outlined as well as some aspects of the measurement of materials contrasts. The presented techniques of nano-scale modification are a dynamical sort of plowing lithography and electrical poling, both performed by means of SFM-probes.

Keywords Contact mechanics · Electrical properties · Interface/interphase · Nanolithography · Polymer-matrix composites (PMC) · Scanning/atomic force microscopy (SFM/AFM) · Thermal diffusivity

List of Abbreviations and Symbols

ABS	Poly(acryl-*co*-styrene)/polybutadiene blend
AFAM	Atomic force acoustic microscopy
AFM	Atomic force microscopy
ATR	Attenuated total reflectance infrared spectroscopy
CDIM	Constant dynamic indentation mode
CFRP	Carbon fibre reinforced polymers
CPM	Colloidal probe microscopy
c-SFM	Conductivity SFM
DC-EFM	Dynamic contact mode electrical force microscopy
DGEBA	Diglycidylether of bisphenol-A
DLC	Diamond-like carbon
DMA	Dynamic-mechanical analysis
DMT	Derjaguin-Muller-Toporow model
DPL	Dynamic plowing lithography
EBL	Electron beam lithography
EDX	Energy dispersive X-ray analysis
EFM	Electrostatic force microscopy
EP	Ethylene-propylene
ESEM	Environmental scanning electron microscope
FDI	Force-distance curve indentation
FEP	Fluoroethylenepropylene
FFM	Friction force microscopy
FMM	Force modulation microscopy
HIPS	High-impact polystyrene
HM	High modulus
HOPG	Highly oriented pyrolitic graphite
IRRAS	Infrared-reflection-absorption spectroscopy
JKR	Johnson-Kendall-Roberts model
LB	Langmuir-Blodgett
MALDI-MS	Matrix-assisted laser desorption/ionisation mass spectrometry

M-D	Maugis-Dugdale model
MD	Molecular-dynamics
MEMS	Micro-electro-mechanical systems
M-LFM	Modulated lateral force microscopy
MPDA	Metaphenylenediamine
MWNT	Multiwall carbon nanotube
oLED	Organic light emitting diodes
P(VDF-TrFE)	Polyvinylidenefluoride/trifluorethylene
PAN	Polyacrylonitrile
PBD	Polybutadiene
PE	Polyethylene
PFM	Pulsed force mode
PMC	Polymer matrix composites
PMMA	Polymethylmethacrylate
PP	Polypropylene
PPS	Polyphenylenesulfide
PR	Phenolic resin
PS	Polystyrene
PVDF	Polyvinylidenedifluoride
PVE	Polyvinylethylene
SAM	Self-assembled monolayers
SCM	Scanning capacitance microscopy
SEM	Scanning electron microscopy
SFA	Surface force apparatus
SFM	Scanning force microscopy
SK(F)M	Scanning Kelvin (force) microscopy
SLAM	Scanning local-acceleration microscopy
SNOM	Scanning near field optical microscopy
SPM	Scanning probe microscopy
SSPM	Scanning surface potential microscopy
SSRM	Scanning spreading resistance microscopy
SThM	Scanning thermal microscopy
STM	Scanning tunnelling microscopy
TEI	Thermoelastic instability
TEM	Transmission electron microscopy
TGS	Triglycine sulfate
TM	Tapping mode
TPE	Thermoplastic elastomeric materials
UFM	Ultrasonic force microscopy
UHV	Ultra high vacuum
UV	Ultraviolet
VM-SFM	Voltage modulated SFM
VUV	Vacuum-ultraviolet
XPS	X-ray photoelectron spectroscopy
a	Contact radius

A	Contact area
A_d	Domain area
A_e	Recovered work; work of elastic deformation
$A_i(h)$	Indenter area function
A_o	Minimum domain area
A_p	Dissipated work; work of plastic deformation; hysteresis of the contact lines
A_r	"Reading" oscillation amplitude
A_w	"Writing" oscillation amplitude
A_{rf}	"Reading" free oscillation amplitude
A_{wf}	"Writing" free oscillation amplitude
b	Effective radius of contact
b^{dyn}	Bending amplitude
B	Amplitude of the oscillations
c	Concentration
c_{th}	Specific heat capacity
C	Charge capacity
γ_c	Critical shear strength
d	Width of the dead zone of the interface height step
d_{33}	Converse piezoelectric constant
D	Dielectric displacement
D_{DPL}, D_{FDI}	Depth of the holes in DPL and in FDI, respectively
$D_n(N)$	Profile of the network structure of the thermoset
D_{th}	Penetration depth of the thermal wave
D_T	Diffusion coefficient
δ	Deformation
δ^{dyn}	Deformation amplitude
δ_x	Lateral displacement
γ	Surface energy
E	Young's modulus
E_c	Coercitive field
E_{el}	Electrical field
ε	Dielectric constant
f_{scan}	Scan rate
F	Force
$F^k{}_{fr}$	Kinetic friction force
$F^s{}_{fr}$	Static friction force
Φ	Phase
G	Shear modulus
G^*	Reduced shear modulus
h	Penetration depth
h_r	Remaining plastic deformation at zero load
H	Hardness
$H(\tau)$	Relaxation-time spectrum
k_c	Elastic constant of the cantilever

k_{eff}	Effective force constant
k_{ts}	Stiffness of the contact
k^{lat}_{ts}	Lateral stiffness of the tip-sample contact
K	Reduced elastic modulus
j_W	Heat current density
J	Creep compliance
I	Electrical current
I_C	Electrical current via mechanical contact
I_T	Electrical tunnelling current
L	Scan velocity
λ	Maugis parameter
λ_{th}	Thermal conductivity
M_{33}	Electric field dependent electrostrictive constant
μ	Friction coefficient
ν	Poisson ratio
p	Contact pressure
P	External load
P_{ad}	Adhesion force
P^{dyn}	Force amplitude in force-modulation mode
P_{el}	Electrical polarisation
P^{r}_{el}	Remanent polarisation
P^{s}_{el}	Spontaneous polarisation
P^{sat}_{el}	Saturation polarisation
Q	Exothermic reaction heat
Q_{33}	Polarisation dependent electrostrictive constant
Q_x	Lateral force
ρ	Mass density
r_1	Radius of curvature of the liquid meniscus
r_2	Radius of the liquid bridge
R	Radius of curvature of the tip
S	Stiffness
S_{bp}	Bulk polymer stiffness
S_w	Amount of stiffness change at the position $N=N_0$
S_H	Side of the hole
σ_c	Yield stress
t_d	Characteristic time of diffusion
t_r	Characteristic time of the crosslinking reaction
T	Temperature
T_C	Curie temperature
T_G	Glass temperature
τ	Characteristic response time
U	Voltage
V	Crack propagation velocity
V_{neg}	Volume of the hole
V_{pos}	Volume of the border-wall

V_r	"Reading" voltage amplitude
V_w	"Writing" voltage amplitude
v_{scan}	Scan velocity
W	W idth of the hole
w	Adhesion energy per unit area
ψ_v	Plasticity factor
z^{dyn}	Modulation amplitude
ζ	Thermal conductance
ω	Frequency
ω_{res}	Resonance frequency of the cantilever

1 Introduction

In scanning force microscopy (SFM) [1] microscopic cantilevers with a sharp tip positioned close to the free end of the cantilever are employed. Alongside with the piezoelectric transducers for high-resolution movement, the cantilever represents one of the core elements of microscopes of the SFM type. If not consisting of silicon nitride (Si_3N_4), which is an insulating ceramic, in the vast majority of cases these cantilevers consist of silicon, which belongs to the class of semiconducting materials. Hence, when investigating surfaces of macromolecular materials by means of SFM, two different classes of materials are involved, namely polymers and inorganic semiconductors. Undoubtedly, these materials are ever-present in our daily lives and are playing more and more an increasing role.

Among others, their great success can be traced back to a common trait, namely a kind of versatility, that is the potential of a material to exhibit wide ranges of different properties which are accessible technically by dedicated but similar fabrication techniques and which can be exploited for manifold applications. For instance, in the case of semiconductors the electrical and optical properties can easily be changed by means of doping techniques. Similarly, the stiffness for tensile loading of polymers can be enhanced significantly by techniques such as blending of two or more polymers complementing one another in their mechanical properties or by stretching. Moreover, electrical and optical properties of polymer materials are increasingly being studied and are paving the way for polymers to be used as active elements, such as organic light emitting diodes (oLED). One trait that obviously seems to be common to the two materials classes, i.e. polymers and semiconductors, as well as to the microscopy technique SFM, is versatility. That is, without significant changes to the basic experimental set-up of the SFM type microscope, many different properties of the surface under investigation can be measured and recorded in the form of an image. Consequently, based on the SFM principle, plenty of different measurement techniques have been developed and have largely driven the widespread dissemination of SFM-type microscopes one may find nowadays in many research labs. Referring to the various existing modifications of the SFM principle as well as to some techniques where microscopic probes are being employed which are completely different from the cantilevers typical for SFM (namely in scanning tunnelling microscopy, STM [2], and often also in scanning near-field optical microscopy SNOM [3]), the more general term scanning probe microscopy (SPM) has been coined. In the particular case of an SFM-type microscope, the probe consists of a cantilever and a tip.

Essentially, the versatility of SFM results from the numerous tip-sample interactions which can be exploited for characterising the surface under investigation. For instance, when operating in immediate mechanical contact between tip and sample and when having applied an electrical voltage between an electrically conductive cantilever and an electrode at the backside of the sample, the current flowing across the tip-sample contact may be used for mapping variations of the electrical conductivity. In an analogue manner, variations of the thermal conductivity of the surface under investigation may be mapped. Whatever the particular set-up, material of the tip surface, and sample material, the force acting between tip and sample can be used for controlling the tip-sample distance. Simultaneously, other interactions such as the electrical tip-sample current can be recorded and used for characterising the sample surface. These kind of imaging techniques reflecting the lateral variations of a particular physical property are often referred to as materials contrasts. They may be used for contrasting different components, differing in their composition or their structure. Similarly, a materials contrast may be used for mapping changes of the measured physical quantity as occurring along a certain direction, but within the same material. Obviously, careful interpretation of the gathered materials contrast requires some knowledge on the tip-sample interaction being exploited. Owing to the operation of the tip in the near-field of the sample, that is within rather small a distance from the sample surface (<100 nm), the magnitude of the effective tip-sample interaction area is affected generally by the surface topography. Consequently, the potential cross-talk from topography via the tip-sample interaction area should always be borne in mind when analysing materials contrast images.

The ability to characterise various physical properties on a highly local scale may provide valuable information not available by means of more conventional microscopy techniques. Whatever the particular issue why a (macromolecular) material is being investigated, imaging its structure is in many cases a straightforward approach and may provide plenty of information on its morphology, texture, the effects of the particular production technique, or the mechanism of failure of a part made up of that material. Structure-property relationships are being studied widely in order to learn how to deduce materials properties from the materials structure, and in what ways the materials properties may be improved by well-aimed modifications to the materials structure. Alongside with the spatial organisation of a material, compositional heterogeneities are being present in most materials and can play a major role for certain properties such as the mechanical strength. Consequently, sampling of concentration profiles or concentration mapping are valuable techniques in materials science. The required spatial resolution can be provided by scanning a focused probe across the area under investigation, e.g. like in scanning electron microscopy (SEM). Since compositional heterogeneities are in many cases related to corresponding variations in local physical properties, the recorded concentration profiles can be translated into property profiles provided that some knowledge of the respective concentration-property-relationship is given. For instance, this knowledge can be gathered from reference measurements performed on materials with a well-defined chemical composition, so-called reference materials. If performing the reference measurements by means of a macroscopic measurement technique, however, it should be borne in mind that the resulting macroscopic concentration-property-relationship does not hold necessarily on the length scale of the concentration gradient of interest. For instance, the macroscopic tensional strength of a material may be impaired significantly by the presence of mesoscopic flaws such as voids, whereas on the local scale the probability of finding flaws within the sampling volume is reduced and the tensional strength can be enhanced. Hence, microscopic measurements are highly desirable which provide information on the physical property of immediate interest.

For instance, when dealing with fibre-reinforced composites, in general, there exists an interphase present between fibre and matrix with mechanical properties deviating from those of the bulk matrix and of the bulk filler material, respectively. The shape of the corresponding radial stiffness profile shows some impact on the stress distribution occurring around the fibre and, hence, on the overall mechanical performance of the composite material [4]. The stiffness profile may arise either from corresponding gradients of the composition or from structural gradients. Instead of profiling these variations and translating them into a stiffness profile by means of reference data or a model, it seems to be more straightforward to map the stiffness variations themselves. Several examples of stiffness mapping as applied to heterogeneous polymer systems are given in Sect. 4. An application of thermal and combined mechanical SFM measurements as performed on a polymer-matrix composite referring to brake pad materials is be elucidated in Sect. 5. The potential benefits which can be taken from electrical materials contrasts as well as particular measurement techniques and their limits are described in Sect. 7.

Rather than aiming to sampling the sample surface in a non-destructive manner, the directness of the tip-sample interactions may be exploited deliberately for modifying the sample surface. Thus, by addressing particular tip-sample interactions promising for inducing surface modifications and by variation of the strength of the tip-sample interaction, the transition from 'reading' to 'writing' can be made. This capability has opened up new routes towards manipulation of materials on the nanoscale. Alongside with the need for small-scale lithography, this potential capability of SPM-based techniques has propelled a now world-wide rush into nanotechnology. Beyond the new tools for nanoscale modification of materials, operation of the SFM in the 'writing regime' may be used for studying the high-strain behaviour of polymers on a highly local scale. As, in general, physical laws holding on the macroscale are implying some statistics, simple scaling down to sampling volumes encompassing only a small number of molecules may lead to erroneous results. 'Small is different' and promising materials properties may be expected on the nanoscale. Moreover, when considering future information storage technologies based on nanoscale modifications of polymer surfaces [5], issues related to the long-term stability of the nanostructures seem to be of paramount interest. In general, it may be stated that the functionality of polymers is being extended to micro- and nanoscale. These kinds of approaches are expected to open up new ways to micro- or nanoscale devices and their future employment will increase even further the present degree of versatility of macromolecular materials.

An SFM based lithography technique by means of mechanical interactions as applied to surfaces of polystyrene (PS) and polymethylmethacrylate (PMMA) is described in Sect. 6. Similarly, some results of SFM-based poling experiments performed on a piezoelectric polymer are elucidated in Sect. 7.6.

In the following Sects. 2 and 3 some fundamentals of SFM are outlined as well as some basics of nanomechanics relevant to tip-sample interactions, respectively.

2 Fundamentals of Scanning Probe Microscopy and Basic Instrumentation

The general principle of scanning probe microscopy (SPM) is to scan a very sharp probe over the surface under investigation. With probe and sample being so close to each other that their mutual interaction occurs via the near-field instead of the far-field, the tip-sample interaction is very local and a high lateral resolution can be achieved. In fact, atomic resolution was accomplished by means of scanning tunnelling microscopy (STM) and scanning force microscopy (SFM) on metals and ionic crystals. Depending on the particular tip-sample interaction being employed, different physical properties of the sample surface can be mapped. In the case of STM, the current tunnelling between a sharpened metal wire and the electrically conductive sample is measured. Owing to the pronounced distance dependence of the tunnelling current, an extraordinary sensitivity for tip-sample distance changes is given. Provided a finely adjustable actuator for movements in direction vertical to the sample surface, the vertical probe position can be adapted to the topography of the sample surface by using the magnitude of the tunnelling current as control variable of a feedback loop. When scanning the STM probe across the sample surface a topographic map can be recorded with the vertical displacements of the probe representing the height changes of the sample surface. In a similar manner to STM, in SFM the mechanical forces acting between a sharp tip and the sample surface are being exploited for recording topography images. Owing to the ever-present van der Waals forces, SFM measurements can be performed on arbitrary materials without special requirements for electrical conductivity. The total tip-sample interaction may also include other forces such as electrostatic or magnetic ones, or solvation forces. The force sensitivity is achieved by means of a microscopic cantilever the tip is attached to. Usually, the bending of the cantilever upon attractive or repulsive tip-sample forces is sensed by means of laser beam reflected off the rear side of the cantilever. The angle of reflection of the focused laser beam changes with the bending of the cantilever. A segmented photodiode may be used for detecting the position changes of the reflected laser beam. Depending on the geometrical and electronic amplification of this detection scheme, which is adapted from the operation principle of a light pointer, a vertical resolution of the order of ~0.01 Å can be achieved. In a similar manner, the torsional flexure of the cantilever due to lateral forces acting on the tip can be detected. This requires sensitivity of the photodiode for positional shifts of the reflected laser beam along a second axis aligned perpendicular to the first one. Hence, fourfold segmented photodiodes are often used for accomplishing that.

Beyond imaging the sample topography, the SFM set-up can be exploited for imaging further sample properties, such as local stiffness, adhesion, friction, or thermal conductivity. In principle, information related to the mechanical properties of the tip-sample contact are accessible without special modifications of the cantilever. The force changes upon approaching the cantilever to the sample surface and the subsequent retracting are generally recorded in the form of force-distance curves and can provide a wealth of information about the particular tip-sample interactions and the mechanical properties of the tip-sample contact [6]. Measuring and analysing force-distance curves is a specialised field and discussing the details is beyond the scope of this article. Essentially, the force-distance curve encompasses an attractive range including the jump-to-contact in the approaching part, a non-linear repulsive range in both the approaching and the retracting part, and a pull-off range in the retracting part. Special-purpose set-ups were devised for probing specific tip-sample interactions like electric, magnetic, thermal or optical ones.

The basic set-up of SFM microscopes is depicted in Fig. 1a. The cantilever is scanned across the surface along lines parallel to the x-axis (axis of fast scanning). Successive scan lines are shifted along the y-axis (axis of slow scanning). Usually, the axis of fast scanning is chosen perpendicular to the length axis of the cantilever. The scan range may range from some nanometres to some tens of micrometres. The tip-sample interaction can be sensed in contact, in intermittent contact, or in non-contact mode. In contact mode the tip touches the sample surface permanently and the overall force on the tip is repulsive. The resulting bending of the cantilever is concave. The repulsive tip-sample interaction changes fast with penetration depth which allows a high z-resolution. During scanning lateral forces act between tip and sample, either due to shear deformation of the tip-sample contact or due to frictional forces. Both of them can be a matter of interest, but the shear forces can result in some modification of the sample surface. In non-contact mode no intimate contact between tip and sample is established but the tip is close enough to the sample surface to sense the vertical gradient of surface

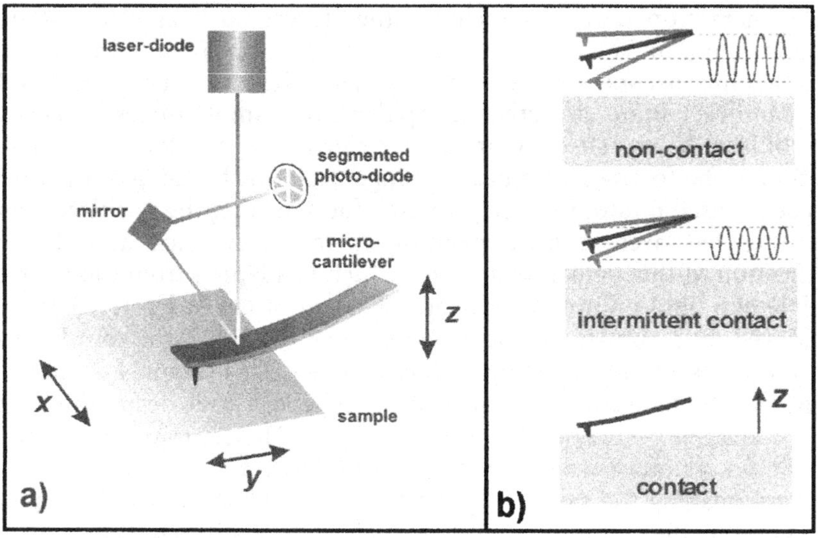

Fig. 1 a Schematic representation of the basic components of a standard SFM detection system. The sample surface under investigation is scanned by means of a very fine tip which is located near the free end of a microscopic cantilever. The vertical forces acting between tip and surface result in a bending of the cantilever which is detected by means of an optical detection system strongly resembling a light pointer. b According to the regime of the force-distance curve where the tip is operated during imaging, three basic operation modes can be distinguished. In non-contact and in intermittent contact operation, the vertical cantilever position is subjected to a periodical displacement modulation with a frequency close to the resonance frequency of the cantilever. In non-contact mode the tip is moving solely in the regime of purely attractive forces, whereas in contact mode the tip-surface forces are repulsive. In intermittent contact mode during each modulation period the tip senses both attractive and repulsive forces with the time of repulsive interaction being only a fraction of the modulation period

forces. The stiffness of the free cantilever is diminished by the vertical gradient of attractive surface forces. The corresponding reduction of the resonance frequency of the cantilever can be exploited for driving the topography feedback. In intermittent contact mode the tip moves both in the attractive and the repulsive regime of the force-distance curve. At the lower point of the vertical displacement modulation of the cantilever base the tip touches the sample surface. In a more figurative way, this mode is also referred to as tapping mode (TM). Pulling-off the tip from the adhesive surface requires a certain minimum amplitude of displacement modulation. The closer the cantilever approaches the sample surface, the more its oscillation amplitude is diminished. Thus, the amplitude signal is generally exploited as input variable to the topography feedback loop. The phase shift of the cantilever oscillation to the driving signal can be related to stiffness and adhesion properties of the surface under investigation.

Extensive research work has been performed in order to characterise the potential and the limits of each of the three basic modes of operation. Since these issues are beyond the scope of this article we may refer to some more dedicated articles and the references given therein [7–15]. The intermittent contact mode has gained great importance for studying thin films of compliant materials and clusters only slightly adhered to the underlying substrate. Since the tip is in contact with the sample surface only for a fraction of the oscillation period (typically, ~6–20 μs) of the cantilever, the scanning-induced shearing forces imposed onto the tip-sample contact are negligible and even the most compliant polymers can be imaged in a non-destructive manner. The interpretation of the phase signal, however, is still a matter of vivid debate. Without claiming completeness, we may refer to some exemplary publications [8, 16–24]. Although the permanent tip-sample contact imposes to some extent lateral forces on the sample surface upon scanning, the contact mode operation is frequently used in studies where the film-substrate adhesion is strong enough, or thick films or even bulk samples are being investigated [25–31].

Mapping a certain area of the sample surface by scanning with the topography feedback being switched on essentially means that the total force acting on the tip is kept rather constant. Presuming that the surface forces acting between tip and sample do not exhibit significant changes, the adjustments of the vertical cantilever position necessary for keeping the preset value of the cantilever bending are equal to the height changes of the scanned topographic profile. Conversely, the scan velocity can be set to zero and the whole force-distance curve can be driven through. For given environmental conditions and a given cantilever, recording the whole force-distance curve provides the maximum amount of information available on a certain site of the surface under investigation. Frequently, the acquisition of force-distance curves is referred to as point spectroscopy. Several approaches have been developed for combining these two opposing acquisition techniques. One obvious approach is to scan the same area several times with the setpoint of the topography feedback being varied successively. However, only a rather limited force range is accessible by this method due to the instabilities of the feedback occurring at small loads. Instead of scanning continuously, the cantilever can successively be moved to different sites of an array and complete force-distance curves can be driven through and recorded on each site of the array. In this way a volume is mapped with each discrete lateral site being represented by a force-distance curve. The lateral translation can be performed either with tip and sample being in intimate contact or at the point of maximum distance between tip and sample. Owing to the fact that no lateral forces are imposed on the tip-sample contact, the latter approach may considered advantageous. In an approach described by Cappella et al. [32], the tip is moved towards the sample surface until a certain preset maximum total force (i.e. concave cantilever bending) is reached. All equidistant points of the force-distance curve are referred to this point. The spatial variation of force-distance curves can be visualised by displaying for each site of the array the force measured at some vertical distance from the reference point. Thus a whole series of that kind of force slice is available with each image representing a certain vertical distance from the point of maximum

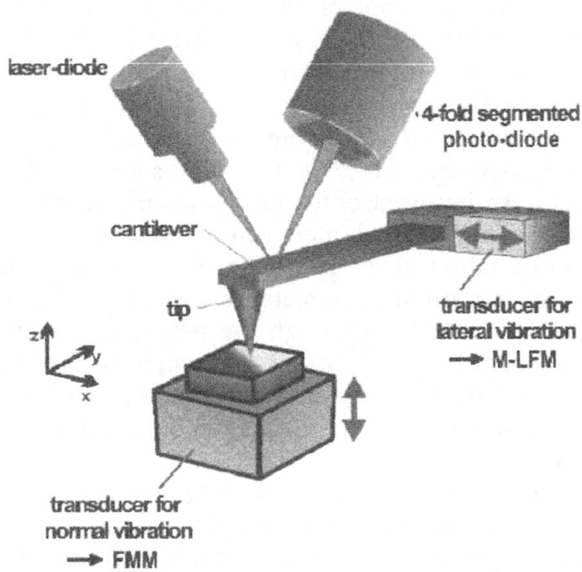

Fig. 2 Schematic representation of the basic detection elements of the scanning force microscope and of the piezoelectric transducers generating the displacement modulations for purposes of dynamic mechanical measurements. The dynamic components of the tip-sample forces resulting from the normal/lateral displacement modulations are detected via the torsion/bending of the microscopic cantilever and the deflection of the laser beam reflected off the rear side of the cantilever. The positional shift of the latter is registered by means of a segmented photo-diode

cantilever bending. However, the rigorous approach of recording complete force-distance curves for an array of surface sites is rather consumptive both in storage space and in time. Whilst storage space is nowadays no more an issue of major concern, the acquisition time still matters. For instance, presuming an acquisition time per force-distance curve of 1 s, the total acquisition time for a 60×60 array takes 1 h. Thus the achievable lateral resolution is rather limited. In a more pragmatic approach the force-distance curve is driven through comparatively rapid while the cantilever is being scanned laterally in a continuous manner (pulsed force mode) [33, 34]. Instead of saving the complete force-distance curve, only some characteristics such as the slope of the contact branch and the pull-off force are being extracted. In a similar approach the lateral translation is performed only when the tip is retracted from the sample surface (jumping mode) [35].

Low-amplitude displacement modulation can be superimposed to either the sample position or the position of the cantilever base; see Fig. 2. Via the stiffness of the cantilever for bending and torsion, the displacement modulation translates to force modulation. More precisely, out-of-plane (i.e. in the z-direction) displacement modulation results in modulated compressive loading of the tip-sample contact [36], whereas in-plane (i.e., in the x- or y-directions) displacement modulation results in modulated shear loading

[37]. Generally, these measurement techniques are called force modulation microscopy (FMM) and modulated lateral force microscopy (M-LFM), respectively. The dynamic response of the cantilever reflects the respective stiffnesses of the tip-sample contact. The highly sensitive lock-in technique can be employed for signal recovery. When using dual-channel lock-in amplifiers, both amplitude and phase of the response signal are available. Moreover, via variation of the modulation frequency, rate-dependent behaviour of the surface material can be investigated. As an alternative to the amplitude of dynamic cantilever deformation, the tip-sample stiffness can be calculated from the shift of the resonance frequencies of the cantilever [38]. Instead of indirect force modulation by means of displacement modulation, direct force modulation was accomplished by applying alternating magnetic fields. Producing a corresponding force on the cantilever requires a magnetisation in the field direction which can be achieved by gluing an appropriately oriented magnetic particle on its rear side [39]. Instead of the particle a magnetic film can be deposited [40]. Especially when operating in a liquid, the systems response is much less obscured by mechanical interferences than in the case of acoustic excitation [40]. By means of a set-up where the SFM cantilever is replaced, Syed Asif et al. [41] implemented a force-modulation technique using a three-plate capacitive load-displacement transducer. The transducer allows both electrostatic force actuation and displacement sensing. Owing to the vertical arrangement of the tip, upon loading there should not occur any appreciable lateral force component. Thus, the amplitude of displacement modulations needs not to be kept small in order to avoid slipping within the tip-sample contact. In SFM-based force modulation experiments, however, the cantilever tilt of ~10–15° with respect to the horizontal causes a lateral force component when loading the tip-sample contact. For minimising this effect, the modulation amplitude should be limited to the nanometre range [42].

When applying the modulation during scanning, amplitude and phase maps can be recorded simultaneously to the image of topography. Alternatively, on selected sites force-distance curves can be driven with simultaneous recording of the dynamic cantilever response signals [43]. With the scanning velocity being set to zero, the tip-sample contact is not affected by sliding. However, when applying in-plane displacement modulations, at high modulation amplitudes sliding of the tip sets in and the lateral forces acting on the tip are due to friction forces rather than shear deformation of the tip-sample contact [44]. Owing to the anisotropy of most of the SFM-cantilevers, the lateral displacement modulation is mostly applied in direction perpendicular to the cantilever length axis, similarly to the scanning motion. Recently, a symmetric probe design has been devised in order to allow measuring lateral forces in arbitrary in-plane directions [45].

FMM was successfully applied in order to image local stiffness variations occurring on surfaces of polymer samples. For instance, Overney et al. [46] and Bar et al. [47] demonstrated by means of FMM that the frictional properties of lubricating monolayers not only depend on their adhesion properties but also on their compression stiffness. Consequently, stiffness should be considered as a major parameter, e.g. when performing friction experiments with systematic variation of the adhesive forces by means of chemically modified tips (chemical force microscopy). Valuable information can be gained from the combined measurement of properties complementing one another. More recently, a combination of the pulsed force mode with FMM was proposed [48]. Thus the low frequency and large amplitude modulation of the vertical cantilever position (responsible for driving through the force-distance curve) is superimposed onto a second vertical modulation of high frequency and low amplitude. In a similar manner pulsed force mode can be combined with lateral vibrations of high frequency and low amplitude, i.e. with M-LFM [48].

Yet another promising combination is studying changes of local mechanical properties upon systematic variation of the sample temperature. Temperature variation was combined with phase imaging in intermittent contact mode [49], FMM [50], or pulsed force microscopy [34, 51]. When heating the total sample, some complications arise due to thermal drift effects. In particular, the stiffness of the cantilever may change upon temperature variation. Instead of applying a macroscopic heating stage, local heating can be achieved by special-purpose micro-cantilevers (scanning thermal microscopy, SThM). A thin thermoresistive filament located at the very end of the micro-cantilever can be used both as a local heating element and a thermal probe [52]. Instead of applying thin filaments, pronounced Joule heating may also be induced at the locus of geometric restrictions of the cantilever cross-section, where the electrical current density is increased [53]. The geometric restrictions can be a part of the cantilever design. When the tip encounters material exhibiting high thermal conductivity, the heat flow into the sample will increase and, consequently, for a given heating current the temperature of the probe itself will decrease. In contrast, by means of a special feedback loop, the probe temperature can be kept constant by adapting appropriately the magnitude of the heating current. As in a standard SFM, the bending of the cantilever upon topographic height changes is used as a control variable of the topography feedback loop. Alternatively to a d.c. heating current, application of a.c. heating currents allows use of the lock-in technique. Beyond the gain in sensitivity, via the frequency of temperature modulation the sensing depth can be varied in a systematic manner [54]. More detailed outlines of the emerging field of SThM can be found in the dedicated reviews by Majumdar [55] and Pollock and Hammiche [56].

Essentially, the sensitivity of FMM for stiffness variations is governed by the ratio of the cantilever stiffness to the stiffness of the tip-sample contact. Moreover, for strong disproportions between these two stiffnesses, the dynamic cantilever response was shown to be affected by frictional forces [57]. For typical cantilevers with bending stiffness, k_c, of ~0.1–50 N/m this condition is easily met when dealing with compliant polymers. In the case of isotactic polypropylene (PP) with the Young's modulus value $E \approx 1.5$ GPa, a tip radius of curvature $R \approx 10$ nm and an imaging force of ≈ 50 nN, the effective tip-sample stiffness is of the order of 30 N/m [26]. Accordingly, stiffness measurements on stiff materials such as metals or ceramics require much stiffer cantilevers. On the other hand, a high cantilever stiffness is equivalent to a low force sensitivity. Especially when investigating materials that are very heterogeneous from the mechanical point of view, conflicting requirements on the cantilever stiffness arise from the presence of both stiff and soft components. For instance, when investigating mechanically heterogeneous materials such as carbon fibre reinforced polymers (CFRP), rather soft cantilevers are necessary for characterising the compliant polymer matrix surface without inducing any unintended surface modification, whereas stiff cantilevers are needed for characterising the stiff fibres. This dilemma can be sidestepped when noting that, for a given cantilever, the range of sufficient stiffness sensitivity can be extended to more stiff materials by operating at frequencies higher than its fundamental resonance frequency. In this regime the cantilever mass behaves like an inertial mass, i.e. the mass of the cantilever cannot move fast enough to keep up with the imposed ultrasonic excitation [58]. Owing to this effect, the cantilever becomes dynamically rigid and even on comparatively stiff materials a soft cantilever can induce elastic surface deformations. For excitation frequencies just above the resonance frequency where the amplitude response decreases like ω^{-2} (where ω denotes the circular frequency of the imposed displacement modulation), FMM was referred to as scanning local-acceleration microscopy (SLAM) [58]. Similar experiments performed above the frequency of the fundamental eigenmode or higher ones are referred to as atomic force acoustic microscopy (AFAM) [59], ultrasonic force microscopy (UFM) [60] or overtone SFM [61]. For operating in the frequency range between ~100 kHz and several MHz and in the acoustic near-field, these techniques may be subsumed under the term near-field acoustic microscopy. At sufficiently high excitation amplitudes the non-linearity of the force-distance curve comes into play and the force averaged over one modulation period is non-vanishing. In a heterodyne approach both the cantilever and the sample were subjected to vertical displacement modulations at adjacent ultrasonic frequencies. Owing to mixing in the tip-sample contact, the cantilever vibrates at the beat (difference) frequency, and the corresponding detection signal can be analysed in terms of amplitude and phase [62]. This approach allows one to monitor phase shifts on the ultrasonic time scale. This feature may be exploited for studying fast relaxation processes.

In the inertial regime the spatial distribution of the cantilever mass is of major importance. Thus, simple spring-mass models where the total cantilever mass is supposed to be concentrated within one point do not hold. Instead, the particular mass-distribution of the cantilever has to be taken into account, e.g. by means of analytical continuum mechanics models [63, 64] or by numerical finite element modelling [53, 61]. As demonstrated by Rabe et al. [63], the continuum mechanical theory of flexural waves of elastic beams applies quite convincingly to SFM-cantilevers made up of single-crystal Si. One of the most striking discrepancies between the continuum mechanics model taking into account the particular shape of the cantilever and the point-mass model is the change of the contact resonance upon increasing stiffness of the tip-sample contact. Whilst the latter predicts diverging contact resonance frequencies, according to the continuum mechanics description the resonance frequencies are bound between the resonance frequencies of the free and of the fully clamped cantilever [63]. Furthermore, in the point-mass model there exists only one eigenmode, namely the resonance of the harmonic oscillator. Consequently, excitations well above the respective resonance frequency are strongly suppressed which is conflicting with the pronounced system's response observable at higher eigenmodes [65]. With increasing amplitude of the ultrasonic excitation, a broader part of the force-distance curve is being sensed, and new effects may be observed owing to its non-linearity. The resonance peaks become increasingly asymmetric, their maxima shift to lower frequencies, and the contact resonances show a hysteresis when the direction of the frequency scan is inverted [66]. Moreover, depending on the adhesion hysteresis the time-averaged force exhibits a discontinuity [60] which is generally referred to as mechanical diode effect. Quantitative evaluation of the resonance spectra in terms of Young's modulus of tip and sample surface requires knowledge of the exact geometry of the cantilever. However, the cross-section of the cantilever beam may not be constant and the tip of the cantilever is quite often not located perfectly symmetric to the central length axis of the beam [59]. As well as the beam shape the shape of the apex of the tip also shows a non-negligible impact on the resonant behaviour. As shown by Dinelli et al. [67], a blunt tip gives no dependence of the contact stiffness on the normal force. Finally, a satisfying description of the dynamics of the tip-sample contact requires the finite lateral stiffness of the tip-sample contact to be taken into account [64, 68].

In general, when performing nanomechanical experiments using a tip-like probe, the overall shape of the tip, its micro-roughness, its chemical composition, and its surface chemistry are as important for the measurement as the topography, roughness, mechanics and surface chemistry of the surface under investigation. Unambiguous characterisation of this surface requires well-defined properties of the probing tip. Thus the conditions of tip preparation should be well-controlled and dedicated techniques are necessary for characterising the tip itself. Essentially, this can be achieved either by employing high-resolution microscopic techniques, such as scanning electron microscopy (SEM) and transmission electron microscopy (TEM), or by calibration procedures. Usually, the latter rely on inert reference samples with quite exactly known properties. Tip characterisation is rendered more and more difficult when the size of the tip reaches nanoscale dimensions. Therefore, despite the reduction in achievable lateral resolution, somewhat blunt tips are often used where the discreteness of matter is considered negligible and the essential parameters such as the radius of curvature of the apex can be determined with sufficient precision. For instance, initially blunt tips may be deliberately worn or covered with a thick coating. In a more rigorous approach, instead of the tip a small sphere-shaped particle attached to the cantilever may be employed for the nanomechanical experiments. This approach is generally referred to as colloidal probe microscopy (CPM) [69]. Micron-size beads of several materials such as glass [70], silica [71, 72], silicon nitride [73], polyethylene [74], polystyrene [75], cross-linked polydimethylsiloxane [76] or cellulose [77, 78] were applied for measuring specific interaction forces. For studying the influence of surfactants or boundary layers, the bead may even be covered with a thin film [71]. Strongly deviating from the sphere-like shape of the attached particle, more recently another kind of nanosize probe technique was proposed, employing a multiwall carbon nanotube (MWNT) [79–81]. When being attached to the end of an SFM-tip with its length axis more or less parallel to that of the tip, the very end of the nanotube may serve as a well-defined nanosized probe (diameter 2–20 nm). The technique is expected to benefit from the extraordinary aspect ratio and the outstanding mechanical properties of carbon nanotubes.

3 Nanomechanics

With increasing surface-to-volume ratio, surface forces play an increasing role, whereas volume forces such as the gravitational force become rather unimportant. In addition to the ever-present van der Waals force, hydrogen bonding forces, capillary forces, or long-range electrostatic forces may give further contributions to the total surface force. For instance, the different surface forces were outlined by Lee [82] and Fröberg et al. [83]. Owing to the adhesive forces, the atoms of the nearby tip and sample may rearrange and the contact area is nonzero even at vanishing external load. When being in contact, the adhesion-driven increase of the contact area between SFM-tip and sample impairs the achievable lateral resolution. Moreover, the SFM-measurement may suffer from fluctuations of the atom positions. Given strong interactions between tip and sample, nanoscale welding can be observed [84]. Specifically, this is the case when bringing uncovered metals into mutual contact. Due to cohesive bonding a connective neck is formed. Upon retracting the tip, the connective neck elongates and finally disrupts at some pull-off force. The neck deforms through a series of structural deformations involving elastic and yielding stages [85]. Molecular dynamics simulations suggest that plastic yielding of nanometre sized contacts under tensile stress involves a series of structural transformations between ordered and disordered states of a small number of atomic layers adjacent to the constriction [86].

The adhesion forces are dramatically reduced when the surface is covered by low-surface energy monolayers such as hydrocarbons and fluorocarbons. Essentially, the coverage with a low surface energy film results in a kind of passivation of the tip-sample contact. In micro-electro-mechanical systems (MEMS) engineering, deposition of low surface energy coatings and chemical surface modification are considered as effective techniques for minimising plastic deformations as well as reducing friction forces and wear [87]. Furthermore, hydrophobisation of the surfaces allows one to alleviate the problem of adsorption of moisture at narrow gaps [88, 89]. Liquid bridging leads to stiction, the unintended sticking of moveable or bendable machine parts to nearby surfaces. Originally, the term stiction was introduced to characterise the failures of hard disk drives in which the available accelerating force is insufficient to overcome local static friction due to adhesion forces among disk, slider and intervening lubricant [90]. In addition to the application of low surface energy coatings, the propensity for stiction may be reduced by minimising the apparent contact area. This can be achieved either by introducing regular microbumps/microdimples or by systematic modification of the microscale surface roughness [88, 89]. Beyond MEMS applications of molecular surface layers, biocompatibilisation of artificial microelectronic interfaces in order to provide reliable operational conditions for biochips under conditions of natural chemical or biological environment is another issue of growing importance [87].

The finite size of nanoscale contacts has strong implications on their mechanical properties. Recalling that plastic deformation in macroscopic crystals usually occurs by motion of dislocations, it is self-evident that the probability of nanoscale plastic flow being affected by dislocations is quite low. The concentration of dislocations in a crystal varies strongly, but even in an annealed single crystal the average separation between two nearby dislocations is small as compared to traditional engineering length scales, namely ~1 μm [91]. In the case of nanoscale mechanical contacts, however, (pre-existing) dislocations are rather unlikely to contribute to plastic deformations and the yield stress, σ_c, may approach the ideal value $G\cdot(2\pi)^{-1}$, where $G=E(2(1+\nu))^{-1}$ is the shear modulus [91] with ν denoting the Poisson ratio, i.e. the negative ratio of the relative deformations as measured in lateral and in normal direction. Similarly, the probability of the measurement to be affected by flaws decreases with decreasing sampling volume. Experimentally, the yield stress was found to be ~1/30 of the Young's modulus, which is ~80 times larger than the macroscopic yield stress [85]. As shown by Bhushan [92], the hardness values decrease with increasing depth of indentation. For a sliding interface exhibiting near-zero friction and wear, surfaces should be free of nanoscratches and contact stresses should be below the hardness of the softer material to minimise plastic deformation. Consistently with the improved hardness, on the micro- to nanoscale the wear rates were observed to be smaller [92]. Beyond the size of the sampling volume, finite size effects may also affect the behaviour of the lubricant. With increasing contact pressure the lubricant is squeezed out of the contact up to several molecular layers. Structure and physical properties of the lubricant molecules are governed by their specific interactions with the confining rigid surfaces rather than by hydrodynamic quantities such as bulk viscosity.

Owing to the predominance of surface forces and the finite size effects, the mechanical behaviour of nanoscale junctions is in many aspects strongly different from that of macroscopic contacts. The improved hardness and wear characteristics are promising for nano- and microscale mechanical devices. In contrast, nanomechanics and nanotribology cannot be predicted by simple down-scaling of the respective laws known from the macroscopic regime. Although the mechanical models neglecting the discrete structure of matter must break down at the atomic scale, continuum mechanics models taking into account surface forces have proved to hold to rather small length scales. Providing analytical expressions, the various classical contact models have found widespread application in micro- and nanoscale mechanical experiments. And being approved on the micrometre scale, it seems quite natural to apply them to some extent to downscaled experiments.

3.1
Fundamentals of Tip-Sample Interactions

Several well-established mechanical contact models are available that are based on concepts from continuum elasticity. In general, the purpose of these models is to describe the deformation of two contacting spheres upon loading. Within the frame of these models, the case of a tip indenting a plane can be considered when setting the radius of curvature of one of the spheres to infinity and approaching the apex of the tip by a sphere of radius R. Despite the fact that these models originate from traditional macroscopic concepts, their use for describing the mechanical contact between SFM-tip and sample surface is rather common and has delivered a considerable number of reasonable results. Taking into account the nano-scale extension of the tip-sample contact, this is somewhat surprising as effects such as spontaneous rearrangements of single atoms are not contained in models that start from averaging over atomic length scales. Results from molecular dynamics simulations, however, corroborated the applicability of classical contact mechanics to nanoscale contacts, at least in a semi-quantitative manner [93]. Via contact mechanics the vital parameter 'contact area' may be estimated, which cannot be directly measured by SFM or related nanoscale indentation techniques. Rather detailed overviews of classical contact mechanics were given in several review articles dealing with nanomechanics [6, 94–96]. This review shall be restricted to the essentials of contact mechanics necessary for grasping the basic ideas and for understanding related SFM experiments.

3.1.1
Normal Loading

Within the widely accepted Hertz-model [97] the mutual deformation of two contacting spheres is described, but without considering attractive surface forces. With increasing surface-to-volume ratio, however, surface forces play an increasing role. Given the same external load, models allowing for surface forces do provide different values of the contact area, the indentation depth and the stored elastic deformation energy. The Johnson-Kendall-Roberts (JKR) model [98] presumes short-range surface forces acting exclusively within the contact area. Owing to these surface forces, the same contact area is achieved for lower values of the external load than in the case of the Hertz model. The area density, w, of the thermodynamic adhesion energy is given by Dupré's equation:

$$w = \gamma = \gamma_1 + \gamma_2 - \gamma_{12} \tag{1}$$

where γ_1 and γ_2 denote the surface energies of the two contacting materials and γ_{12} is the respective interfacial energy. w represents the energy per unit area necessary for separating the two materials and to increase their mutual distance to infinity. In the limit w=0 the JKR equations are identical to those of the Hertz model.

The reduced modulus K of the mechanical contact is given by

$$K = \frac{4}{3}\left(\frac{(1-\nu_1^2)}{E_1} + \frac{(1-\nu_2^2)}{E_2}\right)^{-1} \tag{2}$$

Denoting the radius of curvature of the tip with R and the external load with P, according to JKR theory the contact radius a is given by

$$a^3 = \frac{R}{K}\left\{P + 3\pi wR + \sqrt{6\pi wRP + (3\pi wR)^2}\right\} \tag{3}$$

With P=0 we get $a^3_0=6\pi wR^2K^{-1}$. Hence a is finite even for zero external load P, owing to the adhesion forces. The detachment of tip and sample, the so-called pull-off, requires a tensile force P_{ad} which characterises the adhesion force $P_{ad}=-(3/2)\pi wR$. With δ denoting the deformation of the tip-sample contact as measured in the vertical direction, the corresponding stiffness $k_{ts}=\partial P/\partial\delta$ of the contact is given by

$$k_{ts} = \frac{3}{2}aK \tag{3a}$$

The presumption of short-range surface forces underlying the JKR model is violated when dealing with comparatively stiff materials where the extent of elastic surface deformations is small compared to the range of surface forces. Instead the model according to Derjaguin, Muller and Toporow (DMT) is valid in this case. Following the DMT model [99], surface forces are limited to a ring-shaped area (annulus) enclosing the central contact area where the compressive forces are acting. In the annulus $a<r<c$ the surfaces separate slightly by a distance increasing from zero to a maximal height in the order of the atomic equilibrium spacing. Ruling out deformations due to attractive forces, the resulting deformation profile is of Hertzian nature. The amount of the negative pull-off force is by a factor of 4/3 larger than in the JKR case. As pointed out by Pashley [100], the abrupt decrease of the JKR contact area upon increasing tensional loading implies an according drop in total surface force, finally resulting in the pull-off. Hence, the JKR model predicts separation at a finite contact area, in contradiction to the DMT model where separation occurs at point contact, i.e. the contact with zero deformation. It should be borne in mind that the general presumptions of these contact mechanics models are axisymmetric contacts between isotropic elastic bodies in the absence of shear and in the limit $a<<R$ [101]. The latter condition originates from the Derjaguin approximation of replacing the force interaction between the curved surfaces by an equivalent inter-

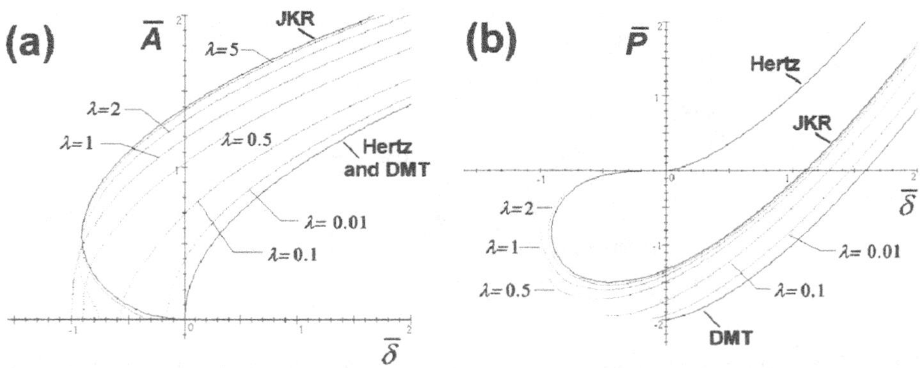

Fig. 3a,b Dependence of the reduced contact area $\bar{A}$and the reduced load$\bar{P}$ on the reduced deformation $\bar{\delta}$, respectively, as calculated by means of the Maugis-Dugdale theory for different values of the transition parameter λ. The JKR, DMT and Hertz limits are indicated. Negative deformations $\bar{\delta}$ occur only in the JKR limit.$\bar{A}$,$\bar{P}$ and $\bar{\delta}$ are dimensionless parameters given by $a(\pi wR^2K^{-1})^{-1/3}$, $P/\pi wR$ and $\delta(\pi^2w^2RK^{-2})^{-1/3}$, respectively. The JKR curves are characterised by nose-like regions; at the extremal point of the nose-like region instability occurs. Adapted from [6]

action between two parallel planar surfaces separated by the same distance. In case of violation of the constraint $a<<R$, the particular shape of the probe needs to be included in a more accurate manner [102].

As shown by Muller et al. [103] and Maugis [104], the transition between the two opposite regimes can be described in terms of a single parameter. Basically, this parameter represents the ratio of the elastic surface deformations to the range of the surface forces. In terms of the Maugis-Dugdale (M-D) model where the Lennard-Jones potential is approximated by a square well potential (Dugdale-potential), the transition parameter λ (which is sometimes also referred to as elasticity parameter) is given by

$$\lambda = 2\sigma_0 \left(\frac{R}{\pi w K^2}\right)^{1/3} \tag{4}$$

where σ_0 denotes the adhesion force acting within the annulus $a<r<c$ and is constant according to the Dugdale type approximation. The M-D model encompasses both the JKR model and the DMT model (Fig. 3).

In the limit $\lambda \rightarrow \infty$ we get the JKR equations, whereas in the opposite limit $\lambda \rightarrow 0$ the DMT equations are obtained. In practice, the JKR regime may be defined by the condition $\lambda>5$ and the DMT regime by $\lambda<0.1$ [105]. In the intermediate range, the transition region, the M-D model has to be applied. The quasi-analytic M-D model, however, does not provide closed formulas and its application for curve-fitting procedures is rather cumbersome due to the fact that a system of equations has to be solved. This is why the development of a direct relation between force and contact deformation is a matter of current efforts [105–107].

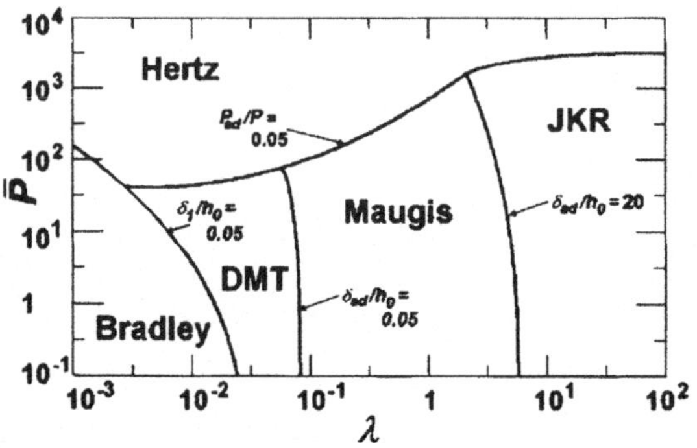

Fig. 4 Map of the validity regimes of the several contact mechanics models (adhesion map). P_{ad}/P denotes the ratio between the adhesive component of the load and the total one. δ_1 is the elastic compression, whereas δ_{ad} is the deformation due to adhesion. h_0 is the effective range of action of adhesive forces (h_0=0.97z_0, whereby z_0 denotes the equilibrium interatomic distance). Adapted from [108]

A map is commonly used for displaying the regimes of validity of the particular contact models (Fig. 4). The coordinates of the map are given by the transition parameter λ as given by Eq. (4) and the ratio $\bar{P}=P/\pi wR$. Here, P denotes the external load and πwR denotes a representative adhesive pull-off force P_{ad}.

The Hertz model applies in the majority of engineering situations where high loads allow for neglecting adhesive forces. In the rigid zone (Bradley regime) all elastic deformation is small compared with the range of the surface forces [108]. The force-deformation relation of the DMT model includes the transition parameter λ which indicates that the DMT model does not encompass the rigid Bradley limit [109]. The DMT model was shown to be valid, in particular when a large meniscus of condensed liquid is present [110, 111].

Similar to Muller et al. [103], several authors performed self-consistent numerical computations of the non-linear equations based on the Lennard-Jones potential instead of dealing with analytical approaches. Greenwood [108, 109] and Feng [112] calculated the dependence of the pull-off force on the transition parameter λ. For fixed γ and R, softer spheres require less force to separate, though they deform more significantly. At the extreme limits of small and large values of λ, however, the pull-off force becomes virtually independent of λ. On a local scale, this is consistent with the prediction of the JKR and the DMT model that the pull-off force is independent of elastic moduli of spheres [112].

The errors made by using the Dugdale approximation may be assessed by comparing with the Lennard-Jones potential [108] which embodies the essential features absent in the Dugdale model, namely the finite repulsive compliance and continuously decreasing attraction. As shown by Johnson and Greenwood [108], the effect of the Dugdale approximation only becomes significant at low values of the transition parameter λ. Similarly, numerical calculations by Barthel [113] indicate that the nature of the interaction has a very moderate influence on the solution. The transition from DMT to JKR is only slightly delayed if a very short-range interaction like a linear one is replaced by a longer-range interaction like van der Waals. The adhesion energy and the geometry of the contact are considered as the dominant parameters.

The low modulus of elastomers generally ensures that $\lambda>80$, so that such materials fall well into the JKR regime [108]. In the case of the surface force apparatus (SFA), the compliance of the glue along with high adhesion results in $\lambda\sim40$, indicating the applicability of the JKR equations. The SFM, however, employs stiff and sharp tips, providing in the case of UHV conditions $0.2\leq\lambda\leq1.1$ [114, 115] and meeting the condition for the transition region. Upon variation of the vapour pressure, the pull-off force may change either due to the adhesion energy w or due to the transition parameter λ [113]. From their measurements with Si_3N_4 tips on high-density polyethylene, Piétrement and Troyon [106] receive from their data evaluation procedure the λ-values 0.045 and 0.143, respectively, measured both under ambient conditions and in vacuum.

The models outlined above are based on the presumption of perfectly elastic behaviour. When dealing with viscoelastic materials such as polymers, however, the rate dependency of the Young's modulus has to be taken into account. Beyond the contact situation of dry viscoelastic materials, some viscous dissipation occurs when tip or sample are covered by thin liquid layers, e.g. thin lubricating films. Given the viscoelastic correspondence principle, stresses and strains of the viscoelastic material can be calculated by replacing the Young's modulus with the creep compliance function $J(t)$, respectively, the stress relaxation function $E(t)$. For very short times ($t\rightarrow0$) the value of $E(t)$ is equal to the instantaneous value E_0, whereas in the opposite limit $t\rightarrow\infty$ the relaxed value E_∞ is attained. The ratio E_0/E_∞ is of the order of $\sim10^4$ for elastomers. Furthermore, adhesion hysteresis is a major feature of indentation experiments performed on viscoelastic materials. The energy necessary for separation is larger than the energy returned when bringing indenter and sample into contact. In general, adhesion hysteresis is governed by the deformation rate. The fundamental contributions to the description of contact mechanics of viscoelastic materials were made by Ting [116], Roberts and Thomas [117], Schapery [118], Maugis and Barquins [119] and Greenwood and Johnson [120]. Beyond the contact radius a the second significant length is the length l along which the surface forces act. The problem is usually described in terms of fracture mechanics and the velocity of the advancing or receding circumferential line enclosing the contact area πa^2 is referred to as crack propagation velocity $V=\mathrm{d}a/\mathrm{d}t$. Thus, the periphery of the contact is viewed as the tip of a crack. Essentially, the deformation rates are given by V/a and V/l. They are the characteristic rates for creep effects and crack effects, respectively. The latter will be dominant when V/l is comparable to the characteristic relaxation rate of the viscoelastic material. For $l<<a$ the highest deformation rates occur in the region of the crack tip. The length l increases with the velocity V. The rate dependence

of the adhesion behaviour is governed by the creep compliance function $J(l/V)$. Glassy behaviour is expected in the limit of rapid tip-sample separation, i.e. in the limit $l/V \rightarrow 0$. At higher loading rates creep effects play an increasing role and their being neglected is no longer justified. A continuum mechanics model including both creep and crack effects was presented by Hui et al. [121]. Essentially, the problem is broken down into an outer problem of viscoelastic contact mechanics and an inner problem of adhesion mechanics. Giri et al. applied this model for analysing their nanoindentation experiments performed on films of tempered styrene-butadiene copolymer latexes [122]. Similar data recorded under low-loading conditions [123], however, were shown to be appropriately described by a model due to Johnson [124] where creep effects are neglected. Along with the loading rate Vakarelski et al. [125] observed some increase of the adhesion force with the holding time at the position of maximal load. Similar to the results of Deruelle et al. [126], as derived from macroscopic measurements, this result suggests that it takes some time for interdiffusion of polymer chains or for establishing bonds across the interface, especially in the case of rough surfaces.

3.1.2
Micro-Indentation Testing

When applying larger indentation depths the presumption $\delta << R$ is no longer justified. In particular, this condition may be violated when using sharp SFM tips which owing to their low radius of curvature, R, easily undergo elastic deformation [127]. Furthermore, considerable penetration may occur either inevitably on very soft materials, such as biopolymers, or rather intentionally if plastic deformation is of major concern. Under these circumstances the shape of the indenter needs to be taken into account more accurately and the contact mechanics should be described in terms of the Sneddon model instead of the Hertz model. Sneddon mechanics describes the deformation of an elastic surface as induced by a totally rigid spherical punch [102]. As illustrated by Heuberger et al. [127], Sneddon mechanics and Hertz mechanics converge in the limit of small deformations and large tip radii. Indeed, blunt tips rather than sharp ones were employed deliberately in several SFM indentation experiments [128, 129]. Moreover, the elastic recovery of the indentation is higher for larger tip radii, due to the hydrostatic nature of the deformation [130]. In contrast, plastic deformation is greater in the case of smaller tip radii.

In a typical indentation experiment the indenter is pressed onto the surface under investigation and the load is successively increased up to a certain maximum load. In the so-called compliance approach both load and indenter displacement are recorded and plotted as a load-displacement curve, the so-called compliance curve. If the experiment is exclusively run in the compressive load regime, the curve is also referred to as the load-penetration curve. Upon loading, elastic deformations occur succeeded by plastic ones. Upon releasing the imposed stress, elastic strain recovers immediately.

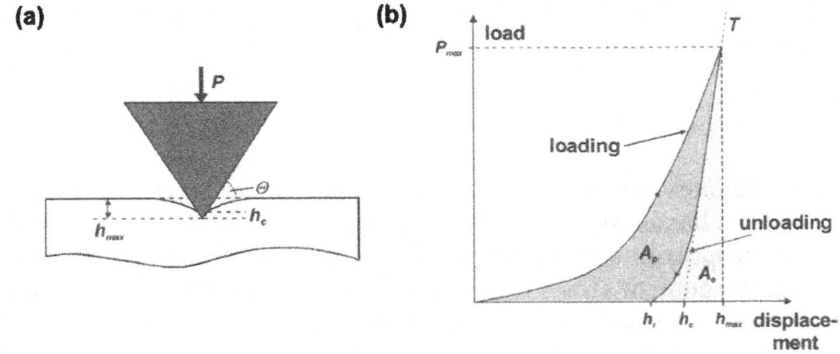

Fig. 5a,b Schematic representation of: **a** the tip-sample contact upon high loading; **b** the according compliance curve. In the case of perfectly plastic response the unloading curve is identical to the vertical line intersecting with the abscissa at h_{max}. In general, some viscoelastic recovery occurs and the residual impression depth h_y is smaller than h_{max}. The difference $h_c - h_y$ represents the extent of viscoelastic recovery. A_p and A_e denote the dissipated and the recovered work, respectively. A_p=0 for perfect elastic behaviour, whereas A_e=0 for perfect plastic behaviour. The viscoelastic-plastic properties of the material may be described by the parameter $A_p(A_p+A_e)^{-1}$. The contact strain increases with the attack angle θ. Adapted from [138]

Additionally, when dealing with viscoelastic materials such as polymers some time-delayed viscoelastic recovery occurs. The residual impression in the sample surface indicates plastic deformation. Both loading and unloading curves are recorded. From the slope of the initial elastic recovery, the contact stiffness can be calculated. Evaluation of the compliance curve in terms of moduli needs knowledge of the area of contact. Hence, tips with rather well-defined geometry are used for performing indentation experiments. The most important indenters are of the Berkovich-type (three-sided pyramidal), Vickers-type (four-sided pyramidal), sphere-like indenters and cone-like indenters. The basic features of compliance curves are depicted in Fig. 5. The most frequently used method of analysis of depth-sensing indentation testing was developed by Oliver and Pharr and is described in great detail in [131]. More recently, a review of the different analysis methods was given by Fischer-Cripps [132].

For any indenter geometry it has been found that the relationship between stiffness S and elastic modulus K may be defined as follows [133]:

$$S|_{h_{max}} = \frac{\partial P}{\partial h}|_{h_{max}} = \frac{3}{2}\frac{\beta}{\sqrt{\pi}}\sqrt{A_{max}}K \tag{5}$$

where A_{max} denotes the projection of the contact area between the indenter and the material at P_{max}, and K denotes the reduced modulus as defined in Eq. (2). For distinguishing from purely elastic deformations, δ, in the case of high-strain indentation experiments exceeding the elastic limit, the penetration depth is referred to as h. β is a parameter dependent on the indenter geometry ($1<\beta<1.034$). The hardness H may be defined by the ratio of P_{max} and the contact area $A(h_c)$:

$$H(h_c) = \frac{P_{max}}{A(h_c)} \tag{6}$$

Here, h_c denotes the vertical distance along which contact is made. For soft metals, the hardness is directly proportional to the flow stress, with the constant of proportionality being ~3. This constant is material-specific and depends on the state of strain underneath the indenter. For a fully plastic indentation, this constant is approximately equal to 2.8 for polymethylmethacrylate (PMMA) [134]. It is worth noting that hardness and elasticity are rather unrelated to each other. Whilst the physical origin of elasticity is the stretching/compression of bonds or it may have an entropic origin as for rubber, plasticity has its origin in the creation and motion of imperfections. Accordingly, the concentration of imperfections shows a rather small impact on the Young's modulus, but it affects the plastic flow stress [91].

The correlation between the cross-sectional area of the indenter and the distance from its tip is given by the so-called indenter area function $A_i(h)$. Usage of $A_i(h)$ is based on the observation that, at peak load, the material conforms to the shape of the indenter to some depth h_c [131]. From Fig. 5 the relation $S=\varepsilon P_{max}(h_{max}-h_c)^{-1}$ is obvious. h_c can be deduced from the load-displacement data, and the projected area of contact can be estimated directly from the area function $A_i(h)$. The constant ε is related to the geometry of the indenter. For perfect Berkovich indenters the area function $A_i(h)$ is given by $A_i(h)=24.5\cdot h^2$. The area function $A_i(h)$ is usually calibrated from indentations on hard and plastic materials such as fused silica [131]. Alternatively, corresponding images of the indenter may be recorded via scanning electron microscopy (SEM) or SFM. Vickers and Berkovich pyramidal indenters are presumed to have point-like tips, whilst usually tip regions are more or less blunted due to fabrication tolerance and wear. These deviations play an increasing role with decreasing indentation depths. Scanning of the indenter tip by means of SFM was used for characterising changes of its shape due to wear [135].

As shown by Oliver and Pharr [131] by careful examination of indentation load-displacement data, the unloading curve can be accurately described by the power law relation $P=m(h-h_r)^n$, where m and n are fitting parameters and h_r denotes the remaining plastic deformation at zero load. The initial unloading slope is found by differentiating analytically this expression and evaluating the derivative at the peak load and the corresponding displacement. The procedure of data evaluation outlined above works well only when the key quantities unloading stiffness, indenter area function and load frame compliance are known with some precision. The latter is important especially for large indentations made in high-modulus materials. Imaging the residual impression is not required when applying the indenter area function $A_i(h)$. However, effects such as piling-up, sinking-in or cracking of material around the indenter can occur which lead to inaccuracies in the evaluation scheme as outlined above. Thus imaging of the residual indent seems to be useful for detecting such irregularities [134, 136]. Again, SEM or SFM may be used for imaging the residual indent. SFM allows the recording of a true topographic map with calibrated height data; however, it should be borne in mind that the SFM topography image is a convolution of the true topography with the shape of the SFM tip. This effect may only be neglected if the tip used for imaging is considerably sharper than the tip used for indenting. Calculating the hardness from the ratio of maximum applied load and the contact area as deduced from the image of the residual impression is a more simple approach than the compliance method. However, this is not applicable for materials exhibiting a considerable amount of viscoelastic recovery as no residual impression will be visible [137]. Micron-indentation measurements on polymers revealed creeping effects, resulting in a rounding of the loading-unloading peak [138]. The rounding can be circumvented by holding the maximum load for a certain time long enough to settle down creeping. Moreover, a significant strain-rate hardening effect was observed (e.g. for polymethylmethacrylate (PMMA)), that is, at higher strain rates the polymer accommodates a lower displacement at a constant maximum applied load [138]. Another effect to be borne in mind is work hardening, either due to high strain rates imposed during the indentation experiment or as a result of stresses imposed during the polishing procedure. Surface oxidation or contamination may lead to deviations of the surface properties from those of the bulk. As the regularity of the indenter-surface contact area is impaired by surface roughness, flat surfaces are desirable for running indentation experiments. Last but not least, indentation size effects are an issue of concern. With increasing sampling volume the probability of encountering flaws rises [136]. When investigating thin films, the depth of indentation should be no more than one-tenth of the film thickness in order to avoid the measurement being influenced by the underlying substrate [139]. Failure mechanisms and wear behaviour of thin coatings are increasingly being studied by means of micro-/nanoindentation and micro-/nanoscale scratching [92, 140].

3.1.3
Shear Loading

The tip-sample contact will deform elastically if a lateral force smaller than the force of limiting friction is applied. In fact, microslip occurs at the contact periphery even for very small lateral forces, and upon increasing lateral forces microslip spreads throughout the interface [141]. Consistently, contact area measurements by Savkoor and Briggs [142] during the static friction phase revealed a decreasing contact size with increasing values of the tangential force. When surmounting the force of limiting friction, the entire contact slips and the tip begins to slide. For lateral forces well below the force of limiting friction, however, microslip effects may be neglected [143]. Presuming a lateral force Q_x acting on the contact, the lateral displacement δ_x between different points in the tip and the sample may be written as

$Q_x=8bG^*\cdot\delta_x=k^{lat}{}_{ts}\cdot\delta_x$ [144]. Here, b denotes the effective radius of contact and G^* the reduced shear modulus of tip and sample:

$$G^* = \left(\frac{2-\nu_1}{G_1}+\frac{2-\nu_2}{G_2}\right)^{-1} \tag{7}$$

$$k_{ts}^{lat} = 8bG^* \tag{8}$$

is the lateral stiffness of the tip-sample contact for small lateral forces. Owing to its definition $b=a+n(c-a)$ with $0<n<1$, the effective radius b over which the frictional traction acts lies between a and c. For the case of UHV conditions, the value $n=0.4$ was shown to be appropriate [141].

In analogy to indentation experiments, measurements of the lateral contact stiffness were used for determining the contact radius [114]. For achieving this, the finite stiffness of tip and cantilever have to be taken into account, which imposes considerable calibration issues. The lateral stiffness of the tip was determined by means of a finite element simulation [143]. As noted by Dedkov [95], the agreement of the experimental friction-load curves of Carpick et al. [115] with the JKR model is rather unexpected when considering the low value of the transition parameter $\lambda(0.2\leq\lambda\leq1.1)$. Further work seems to be necessary in order to clarify the limits of validity of the particular contact mechanics models, especially with regard to nanoscale contacts.

In a dynamic approach the lateral displacement can by modulated, thus allowing the use of highly sensitive lock-in amplifiers [37, 145]. This technique is also referred to as shear modulation. Upon increasing modulation amplitudes the transition from static friction to sliding may readily be observed from the out-of-phase component of the dynamic response signal [44]. When running lateral displacement modulation experiments at MHz frequencies, i.e. well above the fundamental resonance frequency of the cantilever, lift-off of the cantilever was observed [146]. This effect was explained in terms of elastohydrodynamic lubrication, thus taking into account the mechanical behaviour of the fluid film covering tip and sample surface.

Wahl et al. [147] applied SFM-based shear modulation in order to characterise the viscoelastic response of polyvinylethylene (PVE) in terms of the characteristic response time τ. In their mechanical model the viscoelastic component of the contact is described by means of a Voigt model, that is a parallel configuration of a linear spring and a dashpot. Being only dependent on the measured phase shift and on the ratio of measured amplitudes, the calculation of τ did not require knowledge of the cantilever torsional stiffness, thus reducing considerably calibration-related uncertainties. The experiments revealed dependency of τ on frequency and a decrease of τ upon increasing storage time of the PVE sample, which are both characteristics of viscoelastic behaviour.

3.1.4
Sliding Friction

The lateral force necessary to initiate sliding motion (i.e. the static friction force F^{s}_{fr}) is closely related to the normal force. However, the lateral force that must be continually applied to maintain sliding motion at a given velocity, F^{k}_{fr}, was found to be related to the irreversible part of the adhesion energy (hysteresis) as measured during a loading-unloading cycle [148]. The sliding of surfaces relative to each other in or very close to true molecular contact is generally referred to as interfacial sliding. During interfacial sliding, the friction depends critically on the precise distance between the two smooth surfaces, the intermolecular forces between the surfaces, their area of contact, and the sliding velocity [149]. At low loads the frictional force between an asperity and a counter-surface is described by an equation originally proposed by Bowden and Tabor [150]:

$$F^{k}_{fr} = \gamma_c A \tag{9}$$

where A denotes the molecular contact area. γ_c is the critical shear strength at the contacting interface, that is the shear force per unit area necessary for causing slip in the stick-slip regime [151]. The dependence on the contact area is in contradiction to Amonton's law where the friction force is proportional to the load instead of to the contact area ($F^{k}_{fr}=\mu P$). Amonton's law along with its related friction coefficient μ is related to macroscopic friction, that is when sliding occurs between rough surfaces in the presence of wear. As shown by Greenwood [152], the laws of friction arise from the statistics of surface topography. Whilst for single asperity contacts the Bowden and Tabor relationship Eq. (9) seems to be appropriate, Amonton's law is valid when rough surfaces with some distribution of asperity heights and widths slide past each other. Furthermore, when plastic deformations occur upon increasing loading P, the area of contact will grow until the mean contact pressure $p=P/A$ in both materials does not surmount the yield pressure p_y. With $p \approx p_y$, $A \propto P$ and Amonton's law holds with $\mu \approx \gamma_c/p_y$. To sum up, Amonton's law will be valid for multiasperity contacts in relative motion in both the elastic and the plastic regime [153]. In a more general approach the friction force can be thought of as a superposition of the adhesion contribution according to Bowden-Tabor and a load-dependent structural contribution according to Amonton. The superposition exactly holds if the critical shear strength, γ'_c, includes a term αp which is proportional to the pressure p:

$$\gamma'_c = \gamma_c + \alpha p \tag{10}$$

Equation (10) was shown to follow from a thermally activated model of Eyring's type, that is a stress-assisted thermally activated process for viscous flow [154]. For rough surfaces the area of true molecular contact is very small and, hence, the adhesion contribution to the total friction force is usually negligible, as manifested by a zero friction at zero load [148]. Yet in the microscopic limit of the SFM experiment employing a comparatively sharp tip, a finite frictional force was observed at the pull-off point, thus indicating a finite contact area [115]. This finding is in accordance with the JKR theory. From the JKR equation Eq. (3), in the Hertzian limit w=0 we get $A \propto a^2 \propto P^{2/3}$, which along with Eq. (9) means $F^k_{fr} \propto P^{2/3}$. For taking into account adhesive forces, in a simplified approach the external load P may be replaced by the sum of the external load and the adhesive force (Hertz-plus-offset model). As shown by Putman and Kaneko [155], Meyer et al. [156] or Schwarz et al. [153], the 2/3 power law holds rather precisely as long as the shape of the tip apex is close to that of a sphere. Moreover, the observed consistency strongly suggests the validity of contact mechanical theories down to tip radii of only a few nanometres. For specific atomic scale friction models considering lattice structures and particular tip-sample interactions we may refer to dedicated reviews and the references therein [95, 151, 157–159].

When dealing with friction lubricated by a liquid, the regime of hydrodynamic friction should be distinguished from that of boundary lubrication. According to the so-called Stribeck plot, the latter occurs for low values of viscosity or sliding velocity, and high values of the load. Under these conditions the fluid will be squeezed out of the contact area except for some monolayers. While the viscosity is the most important parameter in hydrodynamic lubrication, this parameter is irrelevant for boundary lubrication where the nature of the direct interaction between the solid surfaces and the lubricant molecules is of major importance [91]. As concluded by Carpick and Salmeron [151] from an outline of several SFM friction experiments on model lubricants (such as self-assembled monolayers (SAM) or Langmuir-Blodgett (LB) monolayers), high friction is associated with high adhesion and/or high compliance. Whilst adhesion is dominated by the interaction between chemical groups, compliance is a property of the entire molecule and of the packing of the chains on the substrate [87]. Poor packing allows more energy dissipation modes, such as chain bending, tilting or rotations. In general, longer chain molecules are more densely packed and the energy dissipation modes are sterically hindered [151].

With the surfaces of the sliding counterparts being covered with lubricating films or just organic contamination layers, both the adhesion hysteresis and the friction force exhibit significant changes with temperature and loading rate. The adhesion hysteresis and friction forces were represented by means of a so-called dynamical phase diagram [148]. The underlying rationale is similar to the glass transition known from viscoelastic polymers. In fact, the glass transition is of kinetic rather than of thermodynamic nature, essentially reflecting the temperature dependence of the characteristic relaxation times. The description of the boundary layer in terms of "aggregate states" is convenient but does not refer to any fundamental difference in the thermal equilibrium properties [91]. Rather the classification refers to the ratio of the characteristic relaxation time of the boundary layer to the typical time scale involved in the experiment (Deborah number). Maximum values of adhesion hysteresis and friction are found around a certain chain melting temperature. Instead of temperature, the bell-shaped curve can be plotted as a function of relaxation time or degree of interdiffusion. In the solid-like limit the adhesion hysteresis and friction forces are rather low, owing to the low mutual interpenetration of chains across the interface. In the opposite limit of liquid-like chains the degree of interdiffusion is very high but the system is always close to equilibrium. Again, the adhesion hysteresis and frictional force are low. In the intermediate regime of amorphous chains, however, both the degree of interdiffusion and the relaxation times are significant, resulting in high adhesion hysteresis and friction forces. Increasing the temperature generally shifts the system from the solid-like towards the liquid-like regime. Beyond the dynamical phase transitions of boundary layers, the formation of coherent spatiotemporal structures was proposed as a mechanism for the occurrence of anomalous friction behaviour [160–162]. At the mesoscale coherent structures may appear in molecularly thin films of liquid lubricant when being driven between two sliding plates in the presence of stick-slip boundary conditions. Similar to the phenomenon of stochastic resonances, induced density fluctuations are expected, owing to the cooperation between the external drive and the thermal noise. In particular, at low velocities a reduction in friction may occur with increasing thermal noise or temperature, respectively [161]. Somewhat related, there exist experimental observations of friction reduction upon ultrasonic excitation of the tip-sample contact [146, 163], although the role of the lubricant or the contamination layer in that studies is not yet quite clear. Similar to the application of noise, randomness can be introduced by means of spatial disorder. In agreement with numerical results, friction experiments with krypton (Kr) films on ordered and disordered gold substrates exhibited decreased friction for increased randomness [164].

The rate-dependent behaviour of many lubricating films is also considered as the origin of velocity-dependent stick-slip. Stick-slip transitions are mostly described in terms of freezing-melting transitions of the lubricating layer. Accordingly, the transition from stick to slip, respectively, from static friction F^{s}_{fr} to kinetic friction F^{k}_{fr}, comes along with a transition of the lubricant from the solid-like to the liquid-like regime. At the onset of slip, the intrinsic friction decreases from F^{s}_{fr} in the solid-like state to F^{k}_{fr} in the liquid-like state. The melting process during a slip takes a finite time but appears to be much faster than the freezing process in the stick regime [165]. The analytic model by Carlson and Batista [166] predicts a full range of experimental stick-slip records gathered by means of the surface force apparatus (SFA) [167]. A rate and state law was developed, where the rate variable refers to the sliding velocity, and the state variable captures the shear melting of the lubricant. A boundary line in the dynamical phase diagram is deduced separating the regime of stick-slip motion from that of steady sliding. Similarly, Persson [168] presented an analytic friction law, but for the friction between surfaces lubricated by grafted monolayer films with large separations between the chains, leading to higher chain mobility and lower barriers towards interdiffusion. As compared to the case of dense layers of hydrocarbon fluids or silicon oils having only a weak corrugation of the interaction with the substrate and exhibiting rather abrupt transitions from a solid pinned state to a fluidised state, in the case of the grafted chain molecules having a strong corrugation of the adsorbate-substrate interaction potential, no fluidisation occurs. Instead the sliding occurs at the plane where the grafted chains from the two surfaces meet, and the amplitude of the stick-slip spikes decreases continuously as the spring velocity approaches the critical velocity of steady sliding. As the spring velocity increases, the system will spend less time in the pinned state before the critical stress necessary for initiating the sliding has been reached, resulting in less interpenetration and a smaller static friction force [168].

A particular lubricant playing an important role due to its ever-presence under ambient conditions is water. Adsorption of water molecules from humid air or gas results in thin water films covering the surface of both sample and tip. When measuring force-distance curves, the presence of water in the tip-sample contact area is most obvious from the pull-off force. Since detachment of the tip requires disruption of the water meniscus bridging tip and sample surface, the measured pull-off force reflects mainly the capillary force rather than the adhesion energy between the tip material and the material of the surface spot under investigation. With increasing temperature the pull-off force was found to decrease which indicates some structural changes of the water layers [169]. Along with IR spectroscopy results, the temperature-dependent experiments indicate two different kinds of water layers: beyond the usual liquid layers which are removed up to 100 °C, more strongly bound water layers exist which persist up to 150 °C and under vacuum conditions [169]. The local thickness of the surface water layer depends on the chemical properties of the sample surface. On hydrophobic spots the capillary formation is strongly reduced even at humidities close to saturation [170]. The surface structure and its properties influence the affinity of the water meniscus thus causing a varying capillary force. This may result in changing normal and lateral forces when scanning the tip across the sample surface. The changing shape of the meniscus was nicely demonstrated by means of a hybrid set-up with an SFM-head installed into the vacuum chamber of an environmental scanning electron microscope [171]. However, for symmetric configurations the Kelvin equation was shown to hold for water menisci as small as several hundreds of nanometres [171]. Beyond the humidity and the chemical nature of the sample the transport of water by the SFM-tip also depends on the motion of the tip as revealed by measurements on mica, PMMA and epoxy [172]. Acting as a kind of lubricant and reducing the shear strength of the tip-sample junction, at low loads the action of the water layer is to reduce the friction forces. In the case of NaCl(100) surface, the transition from reversible to hysteretic friction-load behaviour was observed for relative humidities larger than a certain critical value [173]. The qualitative changes of the frictional properties were related to capillary forces coming along with the completion of a water monolayer. Beyond its meaning for SFM measurements performed in mechanical contact or in intermittent contact mode, water deposits may cause instabilities of the feedback when operating in non-contact mode using a compliant cantilever. In general, imaging of water droplets requires operating in non-contact [174] or in intermittent contact mode [175]. Whereas nanodroplets were shown to be clearly mappable, homogeneous thin layers of water are detectable only indirectly, e.g. via height changes or adhesion changes after having modified significantly the environmental conditions. Measurements on hydrophobic graphite surfaces showed that water deposits appear preferentially on scanned areas, thus indicating some tip-induced condensation of water [176]. One promising approach for reducing the capillary forces is just to apply a hydrophobic film to the SFM tip. This was accomplished by Wei et al. [177, 178] by coating their Si_3N_4 tips with silane based molecules. While operating in contact mode they were able to perform friction measurements on water islands condensed on mica. As shown in former studies [179], given a certain threshold water coverage water molecules or clusters are pulled together by capillary action upon tip contact. The resulting monolayer film has an icelike structure induced by the epitaxial interaction with the mica substrate.

4 Stiffness Mapping for the Characterisation of Heterogeneous Polymer Systems

4.1 Characterisation of Crosslinked Polymer Films

In contradiction to UV radiation which is being absorbed efficiently only from particular chemical groups, the so-called chromophores, vacuum- ultraviolet (VUV) radiation can be absorbed from σ-bonding electrons which are present in all kind of organic molecules. Hence, surface modification by means of VUV radiation is not limited to special functional polymers containing chromophores (e.g. resist polymers as used in lithography) but can be applied to commodities such as polyethylene (PE) or polypropylene (PP) [180]. VUV radiation is a typical component of electrical discharges such as low-pressure plasmas [181]. The energy of the photons is sufficient to excite efficiently C–C and C-H σ-bonds of organic compounds [182]. Depending on the absorption spectrum, the electromagnetic radiation can penetrate only some 10 nm into the surface. In case of polyethylene (PE), the penetration depth (95% absorption) of the 120 nm VUV radiation amounts to 52 nm [183]. The absorption results in bond scission and the formation of C=C-double bonds and radicals which undergo secondary reactions, e.g. crosslinking [184]. Considering its typical wavelengths, the VUV radiation lies in between of those of UV-radiation and of X-rays ($\lambda \sim (100–200)$ nm).

Besides spectroscopic techniques such as infrared-reflection-absorption spectroscopy (IRRAS) and X-ray photoelectron spectroscopy (XPS), SFM-based stiffness imaging was applied in order to detect radiation-induced variations of surface stiffness [180]. For that purpose, when exposing the PE-film to the VUV-radiation, the film was covered with a Ni mesh. Thus, the PE-film was partially masked and exposed to the VUV radiation only within the square-shaped holes of the mesh. After having finished that treatment and having removed the mesh, the sample surface was scanned in force modula-

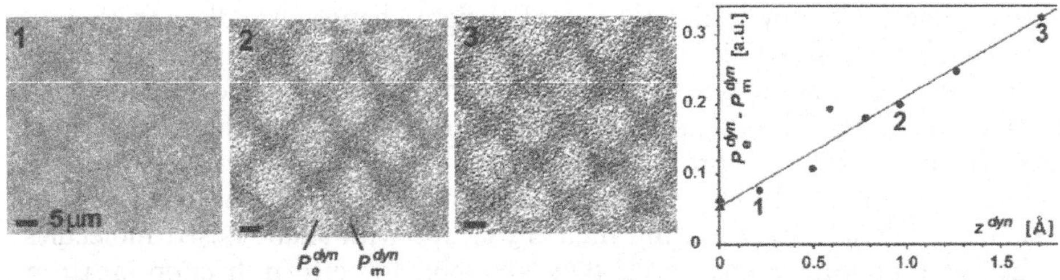

Fig. 6 Amplitude images of a series of FMM measurements performed on a PE-film which was partially masked when exposing it to VUV radiation [185]. In successive measurements, the modulation amplitude z^{dyn} was varied. The irradiated areas exhibit higher values of the FMM amplitude signal (lighter grey tones). Within the given range of modulation amplitudes z^{dyn}, the amplitude difference $P^{dyn}{}_e - P^{dyn}{}_m$ increases linearly with z^{dyn}. The FWHM-widths of the amplitude distributions (see *triangles*) belonging to the exposed and to the masked areas (as deduced from the z^{dyn}=0.2 Å image) are consistent with the ordinate-section of the linear fit curve. Scan width 50.0 μm, scan velocity 15 μm/s, FMM-frequency 45 kHz

tion mode (FMM). Besides some slight differences in surface height as obvious from the topography image, the stiffness image exhibits contrast between the square areas and the partition regions which were masked during the plasma treatment (Fig. 6). Whilst the height changes can be related to some decomposition of the irradiated PE, the increased stiffness within the square areas indicates some crosslinking of the PE chains. The FMM amplitude was measured for a range of modulation amplitudes z^{dyn}. The mean difference $b^{dyn}{}_e - b^{dyn}{}_m$ in the bending amplitude, respectively the corresponding difference $P^{dyn}{}_e - P^{dyn}{}_m$ in the force amplitude, between the exposed and the masked areas was deduced from the histograms of the amplitude images and plotted vs the modulation amplitude z^{dyn}. Within the given range, the stiffness contrast $P^{dyn}{}_e - P^{dyn}{}_m$ rises linearly with the excitation z^{dyn}.

Similar experiments on a PP-film revealed crosslinking as well as decomposition, but with an extent of decomposition much higher than that observed on PE.

4.2
Constant Dynamic Indentation Mode (CDIM) as Demonstrated on a PS/PMMA Composite Sample

When performing FMM measurements by means of displacement modulation, some amplification of the FMM stiffness contrast can be achieved by adapting the modulation amplitude z^{dyn} to the local sample stiffness. From the point of view of contact mechanics this approach makes sense, as the tip-sample contact stiffness k_{ts} is proportional to both the reduced modulus K and to the tip-sample contact radius a (Eq. 3a). With decreasing values of the Young's modulus of the sample material, K decreases whereas a increases. Since on soft materials the value of a is comparatively large, it provides a correspondingly large contribution to k_{ts}. Recalling the correlation $a \propto K^{-1/3}$

from the JKR relation Eq. (3), we get $k_{ts} \propto K^{2/3}$. Thus, the sensitivity $\partial k_{ts}/\partial K$ decreases for increasing values of K.

Hence, if keeping constant the contact radius a during the FMM measurement, the tip-sample contact stiffness k_{ts} would be a direct measure of the reduced modulus K. The static part of a can be controlled be the static load P^{stat}, and the dynamic part of a is given by the modulation amplitude z^{dyn}. In fact, the magnitude of a is not immediately accessible in SFM, but in reasonable approximation the dynamic part of a can be controlled via the corresponding deformation amplitude δ^{dyn} as measured in a direction normal to the sample surface [185]. In the case of displacement modulation, δ^{dyn} is easily accessible as the difference between z^{dyn} and the bending amplitude b^{dyn} of the cantilever ($\delta^{dyn}=z^{dyn}-b^{dyn}$). The decomposition of z^{dyn} into b^{dyn} and δ^{dyn} is valid when working well below the resonance frequency of the cantilever. Both z^{dyn} and b^{dyn} can be measured after having performed appropriate calibration procedures. Given a certain value of b^{dyn}, δ^{dyn} can be adjusted to its set-point value by either increasing or reducing z^{dyn} [186]. On soft sites, the indentation δ of the sample surface by the tip is increased. Consequently, the dynamic component δ^{dyn} is also comparatively large. When operating in the mode of constant dynamic indentation (CDIM), adequate lowering of z^{dyn} is achieved by means of a feedback loop which is working simultaneously to the topography feedback loop [185, 187]. As the response b^{dyn} increases with both z^{dyn} and k_{ts}, the amplitude contrast between stiff and soft sites is increased. This is obvious from the example given in Fig. 7, where FMM amplitude images are displayed as measured with constant value of z^{dyn} (Fig. 7a) and with locally adapted values of z^{dyn} (Fig. 7b,c).

The measurements were performed on a sample consisting of polystyrene (PS) and polymethylmethacrylate (PMMA). After depositing a melt droplet of each polymer on the freshly cleaved surface of an NaCl single crystal and

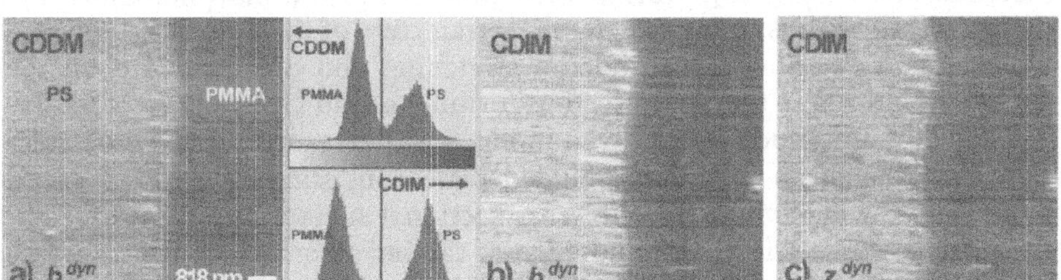

Fig. 7a–c FMM measurements performed on the surface of a polymer composite sample consisting of polystyrene (*left*) and polymethylmethacrylate (*right*) [185]. The surface under investigation was prepared by a replica technique: **a,b** amplitude images of the FMM measurements, driven in constant dynamic displacement mode (CDDM) and in constant dynamic indentation mode (CDIM), respectively. The higher amplitude is measured on PS. In CDI mode, the modulation amplitude z^{dyn} is adapted to the local sample stiffness as visible from the respective z^{dyn}-image given in **c**. The corresponding image of topography (not given) exhibits a maximal corrugation of 23 nm. Scan range 8.18 μm, scan velocity 9.1 μm/s, FMM frequency 45 kHz, cantilever type 'FMR' (supplied by Nanosensors)

after cooling down, the latter was dissolved in deionised water and the replica surface of the polymer composite was investigated. The flatness of the NaCl crystal surface is transferred to the replica surface. Hence, the FMM measurements are not impaired by crosstalk from pronounced topographic features. Furthermore, owing to the well-known location of the two polymer droplets, the FMM contrasts can be attributed unequivocally to either PS or PMMA. The higher FMM amplitude is measured on PS. Despite the rather similar Young's modulus of PS and PMMA, the amplitude images exhibit clear contrasts. Obviously, the amplitude contrast is even increased for the respective image measured in CDI mode (Fig. 7b). The contrast enhancement may also be assessed from the gap of the corresponding histograms. In Fig. 7c the image of modulation amplitude z^{dyn} is given as measured in CDI mode. z^{dyn} is larger on the stiffer phase, namely PS.

4.3 Imaging the Surface Layer of Injection-Moulded Polymer Parts

Injection moulding is a widespread technique for producing polymer parts by extruding polymer melt into a mould. The shape of the mould is defined by the design of the part to be produced. During injection moulding, the polymer melt is subjected to complex deformations and temperature changes. When dealing with blends, such as polypropylene/ethylene-propylene rubber (PP/EP), this complexity increases even further. Consisting of EP rubber particles dispersed in the PP matrix, the PP/EP blend belongs to the group of Thermoplastic Elastomeric (TPE) materials. Owing to the similar chemical structure of PP and EP and to their different physical properties, these blends combine the processing characteristics of plastics at elevated temperatures with the physical properties of conventional elastomers at service temperatures and play increasingly important roles in the polymer materials industry [188]. During melt processing, the morphology of the thermoplastic melt is influenced by the parameters of the flow field. Polymer molecules in the flow front undergo stretching and are deposited on the cooled mould where they solidify, thus forming a skin [189].

The FMM amplitude image given in Fig. 8 was measured on the cross-sectional surface of an injection-moulded plate consisting of a PP/EP blend [190]. The tilt angle between the cross-sectional surface and the surface of the plate is ~75°. The right-hand edge of the image coincides with the line of intersection of both surfaces. The FMM amplitude image reveals an ~14.1 μm wide region exhibiting both a slightly increased FMM amplitude and some structures differing from those farther away from the plate surface. Inspection of the corresponding topography image (not shown) delivers a mean tilt angle $\alpha^x{}_z$ of only ~2.2°. Thus apparent amplitude variations resulting from crosstalk with topography can be widely ruled out and the observed amplitude changes can be attributed to true stiffness changes (Sect. 4.5.1).

Within the skin region, the structures are rather fine and close and run more or less parallel to the right-hand edge of the image, respectively, along

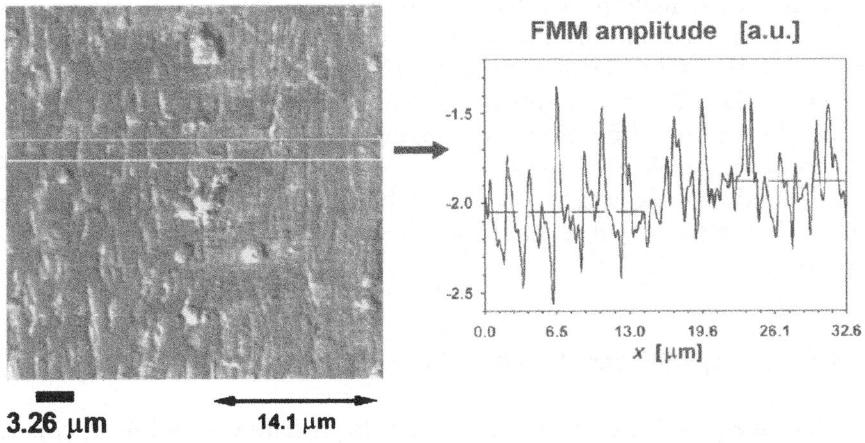

Fig. 8 FMM amplitude image as measured on the cross-sectional surface of an injection-moulded PP/EP blend [190]. The tilt angle between the cross-sectional surface and the surface of the plate resulting from injection-moulding is ≅75°. The right-hand edge of the image is parallel and very close to the surface of the plate. Considering the image, within a ≅14.1 μm wide region, the morphology of the polymer blend and its local stiffness are different from that of the bulk. The skin region exhibits longish structures extended more or less parallel to the sample surface. From the cross-sectional profile the mean changes in the stiffness-related amplitude signal can be deduced. The profile is the average of the 30 successive horizontal lines marked in the amplitude image. The transition region between the two stiffness levels is extended over ≅7.5 μm. Taking into account the tilt of the cross-section with respect to the plate surface, the width of the skin region and of the transition region is ≅13.6 μm and ≅7.2 μm, respectively. Scan range 32.6 μm, scan velocity 25 μm/s, FMM frequency 58.88 kHz, diamond coated 'FMR'-type cantilever (nanosensors). Sample courtesy of Borealis AG, Linz, Austria

the plate surface. From the stiffness profile a transition region ~7.5 μm wide can be identified where the transition between both stiffness levels occurs. Taking into account the tilt of the cross-section with respect to the plate surface, the widths of the skin region and of the transition region are ~13.6 μm and ~7.2 μm, respectively. Both the morphology and the increased local stiffness indicate structural features of the skin region resulting from shear forces acting during injection moulding. The friction between the flowing polymer melt and the comparatively cool mould surface results in strong shear forces which in turn induce some alignment and stretching of the polymer chains. The alignment may come along with an increased mean density which, as well as the stretching, will lead to some stiffening.

Furthermore, the friction forces acting in the flow field can induce phase segregation at the mould surface [189]. As pointed out by Cakmak and Cronin [191], in PP/EP blends with a high content of EP particles even shear amplification phenomena may occur due to the presence of the small rubber particles. The shear amplification results from considerable shear fields occurring in small gaps between rubber particles which in turn are subjected to the macroscopic shear field extended over the whole width of the sample.

Enrichment or depletion of EP rubber particles at the surface of injection-moulded parts may modify the adhesion properties, either via the free surface energy or via the plastic deformation behaviour. As shown by Tomasetti et al. [192], the adherence improvement induced by the presence of EP nodules can be explained by energy dissipation occurring during EP deformation. According to their findings, the magnitude of plastic deformation upon paint debonding is dependent on the depth location of the nodules. This emphasises the important role played by the surface morphology on adhesion properties.

4.4 The Sub-Macroscopic Diffusional Profile of OsO_4 in Polyamide6 (PA6)

Given an unequivocal correlation between local stiffness and local concentration of a chemical diffusing into a material, stiffness mapping may be used for imaging the concentration profile of the penetrating chemical. For instance, in the case of a bulk polymer the local stiffness may be increased by interstitial heavy atoms or by crosslinking. For purposes of selective fixation and staining, treating polymer samples with certain heavy atoms is quite a common procedure when preparing samples for electron microscopy. The desired effect of the heavy atoms is to act as a crosslinking agent for nearby polymer chains. When dealing with semi-crystalline polymers, the incorporation of heavy atoms occurs mainly in regions of reduced packing density, namely the amorphous ones. Due to the presence of the heavy atoms the electron density of the amorphous regions rises over that of the crystalline ones.

Staining of rubber containing polymer blends is achievable by means of osmium tetroxide (OsO_4) which enriches within the rubber inclusions. Beyond the staining effect of the crosslinking reaction with two double bonds of nearby polymer chains, some stiffening of the rubber inclusions occurs which helps to avoid smearing when cutting the material. For example, OsO_4 was used for staining acrylonitrile-butadiene-styrene (ABS) and high-impact polystyrene (HIPS) [193].

A series of six FMM amplitude images is given in Fig. 9a. Starting from a scan area close to the sample surface (left-hand side in Fig. 9a), after each scan the sample was translated towards its core region. Some overlapping of successive scan regions was ensured for stitching the images afterwards. The amplitude of dynamic cantilever bending decays monotonically along the inward direction N. Considering the topographical gradients $\partial z/\partial N$ and $\partial z/\partial T$ indicating the tilting of the surface compared to the plane of scanning, the gradient along the normal **N** is much smaller than that parallel to the sample surface (by the factor ~1/10). The respective angles $\alpha^N{}_z$ and $\alpha^T{}_z$ of inclination are 0.1° and 1.2°, respectively. As the stiffness gradient in the latter direction is negligible, the observed stiffness gradient in the **N** direction may not be ascribed to the inclination of the sample surface. Hence, the stiffness profile P^{dyn} as given in Fig. 9b can be attributed to the concentration profile

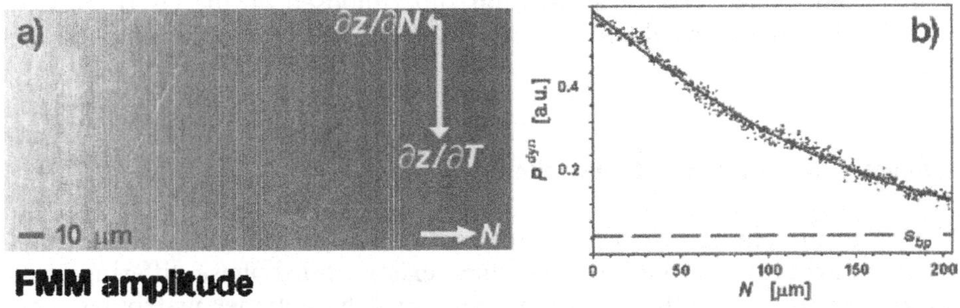

Fig. 9a,b FMM amplitude image of polyamide6 (PA6) subjected to a diffusion experiment using OsO_4 [185]: **a** composition of six images measured successively. After each measurement the scan range was shifted along the N-axis which is pointing inwards the PA6 grain. The lengths of the *arrows* representing the topographical slopes $\partial z/\partial N$ and $\partial z/\partial T$ are proportional to the respective slope angles $\alpha^N{}_z$ and $\alpha^T{}_z$ which amount to 0.11° and 1.16°, respectively. Scan range 100 μm, scan velocity 111.1 μm/s, FMM frequency 130 kHz, cantilever type 'FMR'; **b** amplitude profile as resulting from **a** after averaging over the whole width of the image. The *continuous line* is the result of fitting the amplitude scatter plot by the semi-Gaussian stiffness profile Eq. (12) with a positive prefactor (ν=+1)

of the penetrated OsO_4. The given profile is the mean average of the horizontal line profiles.

In the case of a linear stiffness vs concentration relation, the stiffness profile $S(N)$ reflects the concentration profile without any distortion. With $c(N)$ denoting the concentration profile, $E_s=\kappa_1 c$, and $b^{dyn}=\kappa_2 E_s$, the amplitude profile $b^{dyn}(N)$ may be written as

$$b^{dyn}(N) = b^{dyn}(E_s(c(N))) = \kappa_2 \kappa_1 c(N) \equiv S(N) \tag{11}$$

In a first approximation, the profile $b^{dyn}(N)$ was fitted using a semi-Gaussian diffusion profile with positive prefactor (ν=1)

$$S(N) = S_{bp} + \nu S_w e^{-\frac{1}{2}\left(\frac{N-N_0}{l_c}\right)^2} \tag{12}$$

which is typical of a sample surface covered with a certain amount of diffusing material [194]. S_{bp} denotes the stiffness measured far away from the interface (bulk polymer stiffness) and S_w represents the amount of the stiffness change as measured at the position $N=N_0$. The resulting value of the characteristic length l_c is 203±29 μm; hence the total width of the stiffness profile is $\approx 3l_c$=609 μm. Presuming a coefficient of diffusion D_T being constant during the diffusion process, for a temperature T the characteristic length l_c is given by

$$l_c = \sqrt{2D_T t} \tag{13}$$

With $t\approx$24 h, we get $D_T=2.4\cdot 10^{-9}$ cm^2/s ($T\approx$300 K). It should be noted, however, that the above presumption of a semi-Gaussian diffusion profile and a constant coefficient of diffusion is only a rough approximation for the rather complicated case of a diffusion superimposed on the OsO_4 induced crosslinking reaction which in turn results in the immobilisation of the OsO_4 molecules.

4.5 Interphases in Polymer Based Composites

Owing to the physico-chemical forces acting on the molecules of the polymeric matrix close to the interface, there exists a third phase differing in its properties from both the reinforcing filler and the polymer matrix far away from the interface. Considering the finite thickness of this intermediate layer, it represents a non-vanishing volume. Thus, it is called interphase or less frequently mesophase. The interphase has a major importance for the macroscopic mechanical properties of the composite material, as high stresses may occur close to the filler, depending on the geometry of the filler particles and on the differences in thermal expansion. Due to rapid cooling, the latter may lead to significant initial stresses superimposed to the external ones.

When modelling the mechanical properties of fibre-reinforced polymers, the straightforward approach is presuming an axially symmetric geometry, where the interphase is being represented by a cylindrical volume located between the fibre and the matrix. The corresponding parameters such as Young's modulus, Poisson's ratio or thermal expansion coefficient can be either presumed as being constant (but deviating from the bulk values) or described by certain profiles. For instance, Papanicolaou et al. [195] performed calculations using both a linear and a parabolic profile for Young's modulus and Poisson's ratio. Notwithstanding the specific profile, the general presumption is that the differing values of the parameters are adapted to each other along the interphase. Thus, the radial profiles are supposed to be monotonous.

The experimental verification of that kind of model is difficult to achieve if only measurement techniques are available that deliver signals averaged over regions comparable to or even larger than the total extension of the mechanical profiles. Usually, when dealing with macroscopic measurement techniques, the profiles and the thickness of the interphase cannot be deduced from the measurements without additional presumptions. For the thickness of the interphase of thermoplastic polymer matrices, values smaller than 50 nm were given [196], whereas the interphase of thermosetting polymers was estimated as several hundreds of nanometres [197, 198].

Taking into account the high surface-to-volume ratio of C-fibres (diameter 7–10 μm) or thin glass fibres (diameter 8–15 μm), the relevance of the interphasial volume becomes plausible. For composites with thermosetting matrix and 60 vol.% of fibres, the relative amount of interphasial volume was assessed as $\sim$50% [199]. Along with the high stress intensities typically occurring near the ends of the fibres when the composite material is loaded externally, and along with the micro-cracks frequently observable near the interfaces, the relevance of the interphase is rather obvious. Moreover, the reliability of composite materials is often affected by the environmental conditions. Characteristic parameters such as tensile or impact strength may change significantly when storing the composite under high humidity conditions. In many cases degradation of mechanical properties is attributable to preferential diffusion of water molecules along the interfaces.

4.5.1
The Mesoscopic Stiffness Profile Characterising the Interphase Between Copper(oxide) and Amine-Cured Epoxy

When dealing with polymer matrix composites (PMC) having a thermoplastic matrix, the morphology of the polymer close to the interface with the reinforcing component is expected to deviate from that of the bulk polymer due to interface-related effects such like reduction in segmental mobility, local ordering, entropically induced segregation of molecular weight, or modified crystallisation behaviour.

In the case of PMCs based on thermoelastic polymer, during manufacturing the mixture of curing agent and resin comes into contact with the surface of the reinforcing components, e.g. carbon fibres. The crosslinking reaction of the thermosetting system may be influenced by the presence of filler particles, either chemically via functional groups attached to the filler surface and modifying the crosslinking reaction, or physico-chemically via selective adsorption of molecules of the thermosetting system on the filler surface. As the structure of the resulting network depends on the conditions of the curing reaction such like concentration, temperature, and kinetics, the local mechanical properties which itself are mainly defined by the structure of crosslinks, are expected to reflect the peculiarities of the curing reaction.

By means of infrared spectroscopy, Garton and Daly [200] detected a 200–300 nm thick layer near the polyacrylonitrile (PAN) substrate, where the kinetics of the curing reaction of the epoxy system (epoxy resin Epon828, Shell Chemicals Europe, and curing by means of anhydrides) is different from that of the bulk epoxy system. The observed effect is dependent on the amount of water adsorbed onto the substrate. From the results of their dynamic-mechanical experiments with Epon828/metaphenylenediamine(MPDA) reinforced with C-fibres, Ko et al. [201] conclude that the oligomers of the epoxy resin (polyglycidylether of Bisphenol A) are adsorbed preferentially onto the graphite base plane of the fibre surface. This is associated with a gradient of the relative epoxy/MPDA concentration, which via the local density of crosslinks, shows influence on the shear modulus and the glass temperature.

When studying the morphology of amine-cured epoxies by means of scanning electron microscopy (SEM) during the 1970s, Racich and Koutsky [202] were able to distinguish a 100 nm layer close to the surface of polar materials (glass, copper) differing from that of the bulk epoxy. The nodules (regions where the crosslinking is finished earlier than the macroscopic gelation) occurring near the interface are smaller and their area density is increased. By comparing the morphology near the interface with that near the free surface of the epoxy, they correlate the density and size of the nodules with the local content of curing agent. Accordingly, a smaller size and a higher density of the nodules indicate an increased concentration of the curing agent. Hence, the concentration of the curing agent is increased nearby the surface of the polar materials. Similarly, Fitzer et al. [203] realised a preferred wetting of high modulus (HM) C-fibres by the amine-based curing agent Epikure113 (Shell Chemicals Europe) rather than by the epoxy resin Epikote162 (Shell Chemicals Europe). This tendency was proved for several states of oxidation of the C-fibre.

As pointed out by Palmese and McCullough [204], beyond the thermodynamic driving force for surface-induced segregation of resin and curing agent, the characteristic times of the crosslinking reaction, t_r, and of the diffusion, t_d, have to be taken into account. The increasing viscosity of the mixture of resin and curing agent coming along with the crosslinking counteracts the diffusional motion of the component enriching at the filler surface. In the limit of comparatively short diffusional times, $t_d << t_r$, surface-induced concentration gradients may be adjusted even before the diffusion slows down. In the opposite case of large diffusional times, $t_d >> t_r$, the viscosity increases too fast for establishing a concentration gradient. Thus, building up a chemical gradient and freezing it necessitates comparable times of reaction and diffusion. Owing to the local enrichment of the curing agent and the slowing down of the diffusion, there should occur a complementary region of reduced concentration (Sect. 4.5.3). Via the network structure the concentration gradients should be detectable when studying the morphology or local mechanical properties.

The SFM measurements on thermosetting matrix composites are given in Figs. 10, 11, 12 and 13. With the aim of avoiding any surface modification due to microtoming and polishing the measurements given in Fig. 10 were performed in an alternative way [205]. A 180 nm thick Cu-film was evaporated thermally onto the surface of freshly cleaved mica. By depositing small droplets from the aerosol of an $FeCl_3$ solution onto the free surface of the Cu film, small holes with diameters of the order of several microns were etched into the Cu film (thickness ~180 nm). After cleansing and drying, a stoichiometric mixture of epoxy resin and amine-containing curing agent was poured onto the mica/Cu substrate. The epoxy resin and the curing agent were Epikote828 and EpikureF205 (both supplied by Shell Chemicals Europe), respectively. After curing, the mica sheet was detached from the resulting composite consisting of the holey Cu film with the epoxy layer on top. Afterwards, the bare replica surface of the Cu film was investigated by means of SFM, carried out in force modulation mode (FMM).

As observable from the SFM topography measurements, the surface of the epoxy inside the holes of the copper film exhibits a slight convex curvature, probably due to the surface tension of the epoxy formulation. Hence, the epoxy surface nearby the Cu/epoxy interface was not in contact with the mica substrate when curing the mixture of epoxy resin and curing agent, and any interference on the curing reaction from the presence of the mica can be ruled out. On the other hand, the local slope of the curved epoxy surface is

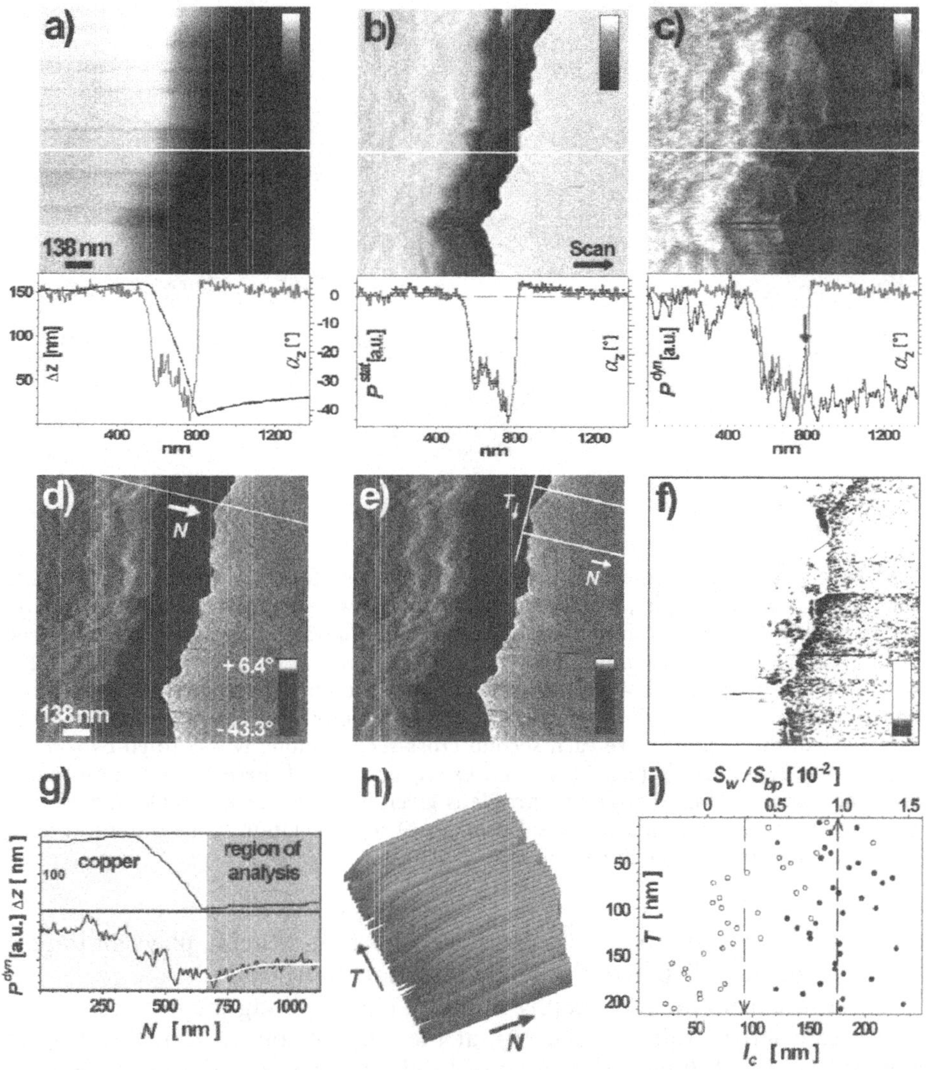

Fig. 10a–i SFM images of the Cu/epoxy interface of a replica sample [205] **a** topography; **b** static load; **c** FMM amplitude (i.e. stiffness). The cross-sections along the white line are shown in the diagrams. The grey-coloured reference line is the corresponding cross-section of the image of tilt angles α^x_z as given in **d**. This image was calculated from the image of topography. The images **e** and **f** are identical with images **b** and **c**, respectively, but are given in enhanced contrast (see *grey tone bars*). The direction normal to the line of the Cu/epoxy interface (T) is denoted by the vector **N**. The *white line* in d indicates the location of the cross-sections shown in **g**, whereas the *white lines* in **e** indicate the boundaries of the region of data analysis defined by the condition $\alpha^x_z<3°$. Within the region of analysis the profile of dynamic force amplitude was fitted by a semi-Gaussian stiffness profile. The *white line* drawn in **g** represents the resulting fit curve. Thirty six cross-sections along the normal direction **N** were analysed. The fit curves and the fit parameters are displayed in **h** and **i**, respectively

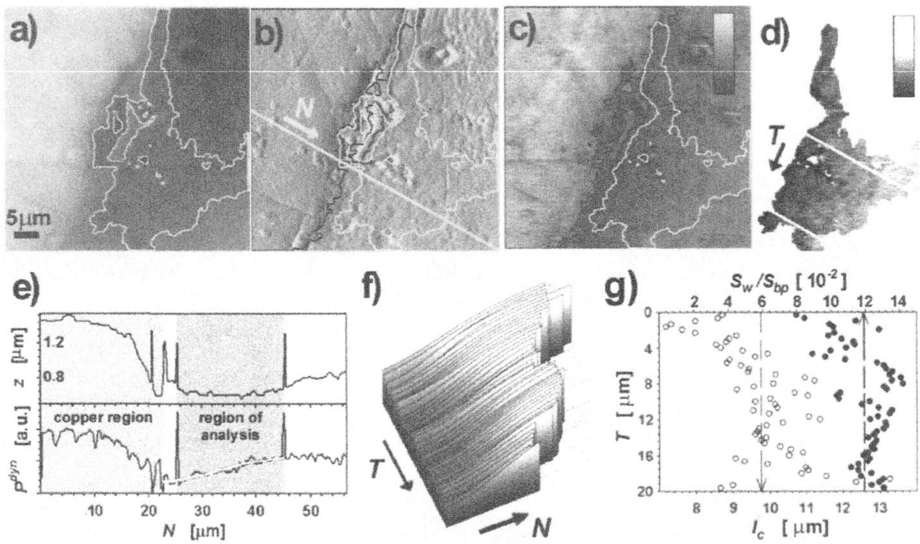

Fig. 11a–g FMM measurement on the cross-sectional surface of a Cu/epoxy composite sample [185]: **a** topography; **b** static normal force; **c** FMM amplitude; **d** region of analysis as marked in **c**, given in enhanced (but linear) contrast of *grey tones*. Scan range 50 μm, scan velocity 55.6 μm/s, maximal corrugation $\Delta z \approx 1.7$ μm, cantilever $k_c \cong 20$–40 N/m, FMM frequency 130 kHz. Within the region of analysis as marked in **c** with a white boundary line, the slope angle α^x_z is <3°. The *white lines* drawn in **d** indicate the boundary of the region where each second cross-section along N was fitted by a semi-Gaussian stiffness profile (Eq. 12, $v=-1$). Exemplary, one of these cross-sections along with the corresponding topography profile is given in **e**. The peaks visible in **e** are due to the marked white boundary layers; **f** pseudo-3D-representation of the fitting curves; **g** display of the values of the fit parameters

rather low, even close to the Cu interface, which, in general, plays an important role for the reliability of the SFM materials contrasts.

As visible from the topography image of Fig. 10, along the fast scan direction (from left to right in Fig. 10), at the edge of the Cu film the drop in height level is ~150 nm. The curve of local topographical slopes as calculated from the topography image ($\alpha^x_z \equiv \arctan(\partial z/\partial x)$) is given in the cross-sectional diagrams of Fig. 10 in a grey colour. At the edge of the Cu film α^x_z decreases to ~−40°, rises steeply in the close vicinity of the Cu/epoxy borderline and slightly decreases within the epoxy region from +4° to ~0°. From the respective cross-section of the static force image it is obvious that the deviations of the normal load from the setpoint value of the topography feedback scale quite exactly with the local slope of the surface under investigation. This correlation explains the pronounced appearance of edges in the image of static load. Along with the scan velocity, the local slope defines the height changes per time unit necessary for adapting the vertical cantilever position to the topographic height differences ($\partial z/\partial t = (\partial z/\partial x) \cdot v_{scan}$). Hereby, v_{scan} denotes the scan velocity.

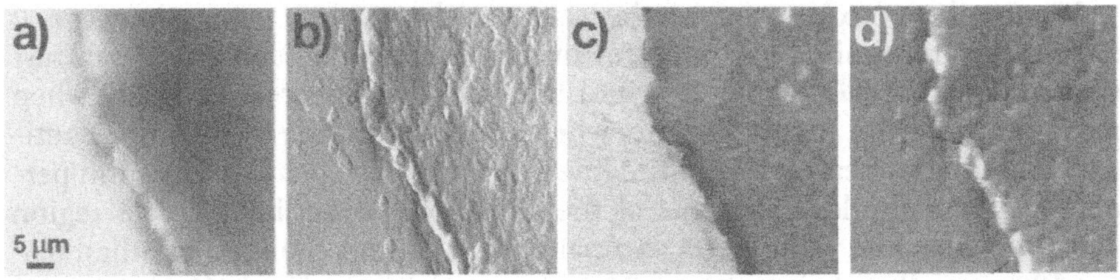

Fig. 12a–d FMM measurement on the cross-sectional surface of a Cu/epoxy composite sample [185]; **a** topography; **b** static normal force; **c** FMM amplitude; **d** FMM phase shift, as measured within the border region between Cu (*left*) and epoxy (*right*). Scan width 50 μm, scan velocity 55.6 μm/s, maximal corrugation $\Delta z \approx 1.6$ μm, cantilever $k_c \cong 20$–40 N/m, FMM frequency 130 kHz. A bulge-like structure along the Cu/epoxy borderline is visible which may not be attributed to the Cu component. Epoxy resin/curing agent: L180/H181 (both supplied by Scheufler)

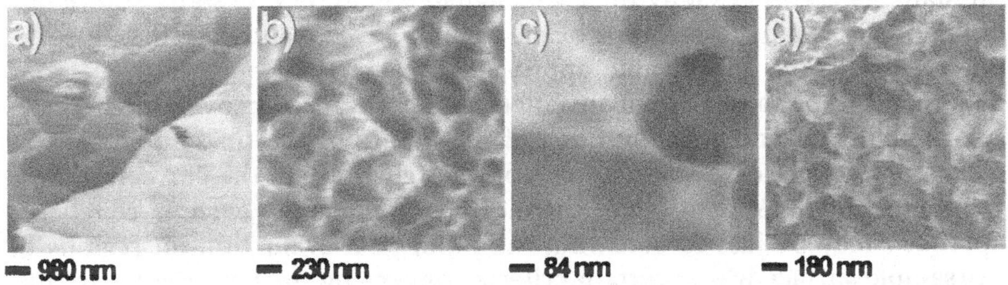

Fig. 13a–d Images of FMM amplitude as measured within the boundary region of Cu and epoxy (**a** and **d**), and far away from the Cu edge (**b** and **c**) [185]. From the amplitude images **a**, **b** and **c** globular structures with reduced local stiffness are recognisable. These need to be distinguished from the structures visible from **d** which exhibit more diffuse shapes and correspond to vein-like structures only visible from the static force image (not shown) belonging to **d**. Hence, the structures of **d** may be attributed to plastic deformations. Epoxy resin/curing agent: L180/H181 (both supplied by Scheufler)

Generally speaking, in regions with high local slopes and slope changes, the geometry of the tip-sample contact is ill-defined, both concerning the size of the tip-sample contact area and its position on the very end of the tip. The tip may contact the sample surface more with its flank rather than with its apex and considerable lateral forces may occur which will induce some severe torsion of the cantilever. As the signals related to the modulation modes (FMM or M-LFM) are strongly related to the tip-sample contact area, they reflect more the topographic features than true changes in materials properties.

On average, the image of dynamic cantilever bending given in Fig. 10c is darker in the region of the steep edge than on the rather flat surface of the Cu film but still much brighter than on the epoxy surface. Hence, the steep edge can be attributed to the Cu film. The peak (indicated with an arrow) in the amplitude signal coincides more or less with the maximal change of surface slope, $\partial\alpha^x{}_z/\partial x$.

Within the epoxide region, the slope angle α^x_z has values between 0 and 4°. As recognisable from the images depicted in enhanced contrast (Fig. 10d–f), the slope angle α^x_z and the static load decrease slightly when moving away from the borderline, whereas the amplitude of dynamic cantilever bending increases. The white lines drawn in Fig. 10e in a direction perpendicular to the local tangent of the Cu/epoxy borderline mark a region where the absolute value of α^x_z is smaller than 3° and changes less than 0.6° per 100 nm (when being at least 10 nm away from the borderline). The dynamic amplitude signal, however, exhibits a monotonic increase extended over ~300 nm which may be fitted by a semi-Gaussian profile.

Considering the lateral extension of the rising gradient of P^{dyn}, the low slope angle α^x_z of the topography, and the low changes of α^x_z, artefacts related to topography can be ruled out and the observed changes in P^{dyn} can be attributed to true stiffness changes of the epoxy.

As taking into account in what way variations in the contact area and in the static load affect exactly the FMM signals would require some knowledge of the precise shape of the apex of the tip and of the features of the topography feedback, the conceivable approach of correcting the measured FMM data seems to be rather an inconvenient undertaking. Moreover, the necessity of making several presumptions of the nature of the tip-sample contact mechanics would leave some doubts on the reliability of the correction procedure for the FMM data. Thus, the more convenient and even more reliable approach may be just to constrain the FMM data evaluation on regions of the sample surface where certain criteria concerning the topography are fulfilled. As illustrated above by considering the cross-sections depicted in Fig. 10, the local surface slope is an important and easily accessible parameter for defining the region of analysis.

To accept this, it is useful to realise that the maximal changes of the surface slope α^x_z the topography feedback loop has to cope with are bound to the maximal amount of α^x_z, $|\alpha^x_{z,\max}|$. If the slope changes $|\Delta\, \alpha^x_z|$ per time unit Δt shall be limited to a fraction $q_1 \cdot |\alpha^x_{z,\max}|$ when evaluating the slope on a length scale given by $\Delta x = q_2 \cdot L$, the slope change per time unit can be written as

$$\frac{|\Delta\alpha^x_z|}{\Delta t} = \frac{|\Delta\alpha^x_z|}{\Delta x} \cdot v_{scan} \leq \frac{q_1 \left|\alpha^x_{z,\max}\right|}{q_2 L} \cdot v_{scan} = \frac{q_1}{q_2} \cdot \left|\alpha^x_{z,\max}\right| \cdot f_{scan} \tag{14}$$

Hereby, v_{scan} denotes the scan velocity, L the scan range and f_{scan} the scan rate. For averaging out slope variations due to surface roughness, the local slope angle α^x_z should be averaged over a suitable length $\Delta x = q_2 \cdot L$. As a rule of thumb, the value of Δx should be no larger than ~1/5 of the (a priori unknown) characteristic length l_c of the stiffness profile. On the other hand, in order to provide a sufficient number of datapoints for fitting the interphasial gradients, the scan range L should be no larger than ~10 l_c. As the maximal change of α^x_z is $2|\alpha^x_{z,\max}|$ when moving from one averaging length Δx to the next one, the maximal value of q_1 is 2. Hence, with $q_2 = 1/50$, the ratio q_1/q_2 is 100. With $|\alpha^x_{z,\max}| = 3°$ and $f_{scan} = 1$ Hz, the necessary slope angle change per time unit, $|\alpha^x_z|/\Delta t$, is smaller than 300°/s. This value is rather low and should be achievable by the topography feedback-loop, even under conditions of mediocre tuning of the feedback parameters. In the case of a slight global surface inclination, which should generally be minimised as far as possible, and solely positive or negative values of the local surface slopes, α^x_z may only change from $|\alpha^x_{z,\max}|$ to 0 within of the distance $2\Delta x$; thus the upper bound for q_l is 1. At small scan ranges L and rather smooth topographic features the value 0.5 for q_l seems to be more realistic and, presuming the same performance of the topography feedback-loop, $|\alpha^x_{z,\max}|$ may be set to 12°.

The conditions $|\alpha^x_{z,\max}|=3°$ and $|\alpha^x_{z,\max}|=12°$, respectively, can be used as a quantitative, easily accessible criterion for defining the region of analysis. In the following, these criteria shall be denoted as 'topography criterion for large scan ranges' and 'topography criterion for small scan ranges', respectively.

In Fig. 10g the cross-sections of the image of topography and the FMM amplitude images are given. The cross-sections are along the white line as marked in Fig. 10d. Similarly to the above case of the stiffness gradient in PA6 (Sect. 4.4)), the FMM amplitude was fitted using a semi-Gaussian profile (Eq. 12), but with a negative prefactor ($\nu=-1$). The value of N_0 is given by the position of the borderline separating the Cu region from the epoxy region. The bulk value S_{bp} of the epoxy stiffness was determined by averaging the tails of all evaluated stiffness profiles. The remaining two parameters S_w and l_c served for adjusting the stiffness profile Eq. (12) to the FMM amplitude profiles. In Fig. 10i the values of S_w and l_c resulting from the fits are plotted vs the lateral coordinate T. Here, the offset suppressed in the images for contrast purposes was taken into account. A pseudo-3D representation of the fit curves is given in Fig. 10h. Herein, the N_0 values were set to zero. The mean value of the relative stiffness change at $N=N_0$, S_w/S_{bp}, is $(9.7\pm2.4)\cdot10^{-3}$. Presuming a value of 3 GPa for S_{bp} and linear $P^{dyn}(E_s)$-characteristics, the stiffness reduction $S_{bp}-S_w$ at $N=N_0$ amounts to 29 MPa. The characteristic length varies between 30 nm and 170 nm; its mean value is 93 nm. With $t\approx10$ min and $T\approx323$ K, Eq. (13) delivers a value of $7.2\cdot10^{-14}$ cm^2/s for the diffusion coefficient D_T.

The semi-Gaussian stiffness profile indicates diffusion processes occurring before the gelation of the reactive mixture of epoxy resin and curing agent. The diffusion processes are expected to modify the local structure of chemical crosslinks. Due to its finite width of $\sim3l_c=280$ nm, the gradient zone has to be considered as a volume rather than as an area and, hence, it deserves to be called interphase.

Alternatively to the above described preparation branch employing Cu films deposited onto mica, Cu/epoxy composites were microtomed and the resulting cross-sectional surfaces investigated by means of FMM. The direction of cutting was perpendicular to the Cu/epoxy interface.

The FMM images and the corresponding data evaluation of a measurement performed on a respective surface are displayed in Fig. 11. Again, the FMM amplitude measured on epoxy close to the Cu/epoxy borderline is reduced as compared to that measured on bulk epoxy. However, in contradiction to Fig. 10, the mean width of the gradient zone is ~30 μm. Hence, it is more than 100 times wider. Working on the basis of Eq. (12) with $\nu=-1$ delivers a mean value of 9.8 μm for the characteristic length l_c and 0.12 for the relative stiffness change S_w/S_{bp}.

It should be noted that, due to the topographic height step often occurring at the borderline between the polymer and the stiff reinforcing component, when dealing with surfaces generated by microtoming, the region of analysis as defined by the above defined criteria based on the topographic slope angle may be too close to the height step. After having scanned across a pronounced height step, the topography feedback may not yet have settled down, even if the tip is already moving within a region where the slope-based condition is met. Hence, additionally the static force signal should be inspected in those cases in order to justify the boundaries of the region of analysis that are close to a step. Thus, there may exist a gap between the reinforcement/polymer boundary where serious analysis of stiffness data is not acceptable, just like in Fig. 11. In these cases possible deviations of the stiffness profile from the trend fitted within the region of analysis will not be noticed. When considering Fig. 11, even within the upper, peninsula-like part of the region of analysis very close to the Cu/epoxy borderline, no increase of FMM amplitude is observable, so that qualitative deviations from the semi-Gaussian profiles can be excluded.

The topography of cross-sectional surfaces produced by cutting and polishing usually reflects the differences in mechanical properties of its heterogeneities. In general, this is due to the fact that the rate of material removal of hard components like fibres and metals is smaller than that of soft polymeric components. Easily deformable components like rubber inclusions, however, tend to be more compressed rather than abraded and their surfaces are quite often above that of the surrounding material of the final cross-section.

In Fig. 12 the FMM images measured on the cross-section of a Cu/epoxy composite are displayed exhibiting an eye-catching bulge next to the Cu/epoxy borderline. The bulge is 3.8±0.6 μm wide. Though the FMM signals measured on the bulge are affected by its strong curvature, the pronounced reduction in amplitude and increase in phase indicate that the bulge material does not consist of Cu. Hence, the bulge may be identified with epoxy of reduced stiffness which was abraded with a rate lower than that of the bulk epoxy.

Taking into account the offset suppressed when scaling the grey tone gradation curve of the FMM amplitude image, the resulting stiffness ratio S_w/S_{bp} is ~0.06 which is half the value of the stiffness ratio deduced from Fig. 11. Whether the bulge-like topography of the interphase may be traced back to its structural properties or whether it is just an effect of the peculiarities of the preparation procedure is difficult to clarify as close-view SFM measurements on the bulge suffer from its strong curvature.

From scans with smaller scan ranges globular-shaped entities can be observed on the epoxy which are rather pronounced, especially in the FMM amplitude images. That kind of heterogeneity could be found both next to the Cu/epoxy borderline and far away from it. Characteristic FMM amplitude images are given in Fig. 13.

In the amplitude image in Fig. 13a granular structures next to the borderline are visible with a total width ranging from 2.5 to 5.6 μm, as measured along the line perpendicular to the Cu/epoxy borderline. This is similar to the interphase from Fig. 12. The globules detectable on bulk epoxy exhibit diameters within the range of 150 and 460 nm; see Fig. 13b,c. Hence, they are smaller than the granules of Fig. 13a that are extended over $\sim$1.5 to 3.9 μm.

The dark regions obvious from Fig. 13d are different from those of Fig. 13a–c in the sense that they are a kind of blurred and that the corresponding static force image exhibits fine vein-like structures indicating plastic deformations probably resulting from shear forces acting during cutting and polishing. Reduced local stiffness may result from the plastic deformations, but it may also be the reason for them. Since the image at Fig. 13d was measured using a cantilever of stiffness lower than that of the cantilever employed when measuring the images at Fig. 13a–c (2.8–5.7 N/m vs 20–40 N/m), it seems unreasonable to attribute the plastic deformations to scan-induced surface modifications.

4.5.2
The Microscopic Stiffness Profile Characterising the Interphase Between C-Fibres and Polyphenylenesulfide (PPS)

As elucidated above, height steps between the surface of filler particles and the polymer matrix are a major hindrance for detecting the interphase by means of SFM. In the opposite case of extremely flat surfaces there arises the problem of identifying the filler/polymer borderline unequivocally. For that purpose a reference signal is desirable from which the location of the borderline may be deduced in a precise manner. When dealing with electrically conductive filler particles providing a continuous current path, the electrical current flowing through the tip-sample contact can serve as an appropriate reference signal. This is just the case when investigating C-fibres embedded in a polymer matrix, with the continuous fibres being contacted electrically at the bottom of the sample and a voltage being applied between sample and conductive cantilever. Owing to the electrical conductivity of the C-fibres and the lack of an isolating oxide layer on the surface of the C-fibres, a pronounced drop in the current signal is easily observable when scanning from C-fibre to polyphenylenesulfide (PPS).

Topography, current amplitude and FMM amplitude images of a respective measurement are displayed in Fig. 14. Scan range is 500 nm and the pixel resolution is 1 nm/pixel. The cross-sectional curves along the line AB are given in the graphs of Fig. 14a–c. In each case, the given cross-sections resulted from averaging over the ten parallel profiles framed by the rectangle AB. For reasons of comparison, the shape of the used SFM tip as deduced from the SEM micrograph given in Fig. 14f is marked in the graph of Fig. 14a.

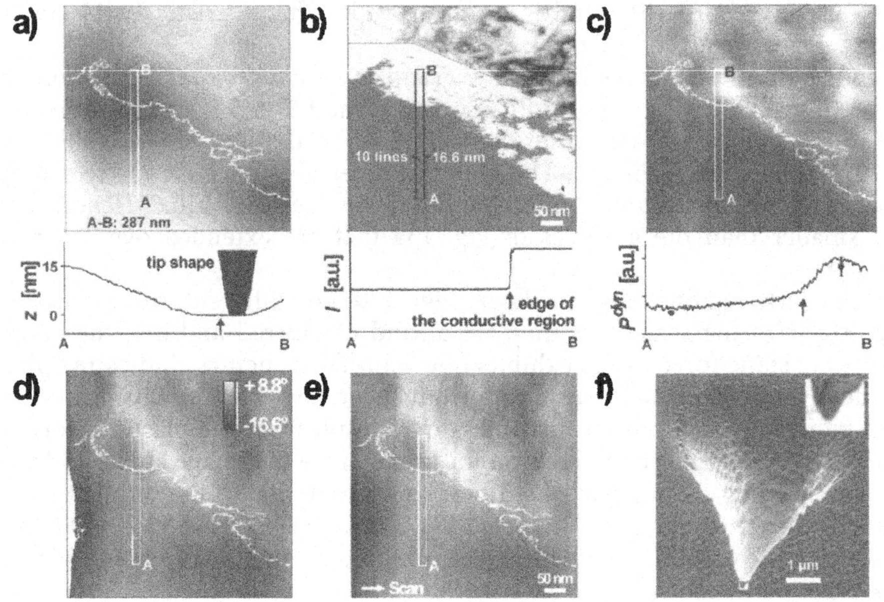

Fig. 14a–e Simultaneously measured SFM contrasts on the cross-section of a C-fibre/PPS composite sample [186]; **a** topography; **b** amplitude signal of the electrical tip-sample current; **c** FMM amplitude. The electrical conductive and stiff fibre (*upper right image corner*) can be identified from the high amplitude signals. The borderline as deduced from the pronounced drop of the current signal is marked in the images **a**, **c**, **d** and **e** by means of a *white line*. Scan width 500 nm, scan velocity 500 nm/s, maximal corrugation $\Delta z \approx 35$ nm, cantilever (supplier NT-MDT, Moscow) $k_c \cong 2.5...6.0$ N/m, tip height 7 μm, tip opening angle <20°, current measurement frequency $\cong$42.61 kHz, FMM frequency $\cong$74.89 kHz. The tip shape as deduced from an SEM micrograph **f** is drawn in the cross-section $\overline{AB}$ of the topography image; **d** image of the slope angle α^x_z as calculated from **a**. Pixels with $|\alpha^x_z|>11.3°$ are given in white colour; **e** image of static normal force

From Fig. 14b,c it is obvious that the edge of the electrically conductive and stiff C-fibre is running from the top left to the lower right of the images. The dark lower part of the amplitude images has to be attributed to the isolating and soft PPS. The borderline as deduced from the electrical amplitude image is marked as a white line in the images given in Fig. 14a,c. From the topography image displaying a maximal corrugation of no more than 35 nm, no valuable information on the location of the C-fibre/PPS borderline can be taken. The upper part of Fig. 14b is represented on a logarithmic scale in order to visualise the broad range of current amplitude values measured on the C-fibre. This may be due either to inhomogeneities of the fibrous material or to variations of the tip-sample contact area.

The mean radius of curvature of the apex of the SFM tip as deduced from the SEM micrograph (Fig. 14f) is ~23 nm. Presuming a spherical shape of the apex and applying JKR contact mechanics (Sect. 3.1.1, Eq. 3), the tip-sample contact radius was estimated to be less than 11 nm (P^{stat}=90 nN,

E(PPS)=3.31 GPa, γ=30 mJ/m^2). Relating this to the lateral lengths along which the topography changes occur, this value is rather small and deviations of the measured topography from the true one due to convolution effects seem to be negligible. Since the characteristic length scale of topography variations is of the order of the scan range, the above defined topography criterion for small scan ranges may be applied, constraining the region of analysis to tilt angles $\alpha^x{}_z$ smaller than 12°. This condition is violated only within a small region at the very left of the images, which was marked with white colour (Fig. 14d).

From the pronounced drop of the current amplitude occurring when scanning from fibrous material to PPS, the position of the fibre/matrix borderline can be deduced with a precision of ±1.7 nm. Within the corresponding graphs the lateral position of the current drop is marked by means of a vertical arrow. As visible from the FMM amplitude image, the force amplitude P^{dyn} rises already several tens of nanometres ahead of the current drop. The local increase of the polymer stiffness when approaching the C-fibre from the PPS indicates the existence of an polymeric interphase with properties deviating from those of the bulk PPS. The black circular dots drawn in the graph of Fig. 14c represent the mean values of P^{dyn} as calculated from at least 12,780 pixels measured on bulk PPS and on bulk C-fibre, respectively. In the case of the bulk PPS, the standard variation is smaller than the radius of the dot.

When using the current amplitude as a reference signal for defining the fibre edge, it should be noted that due to the existence of tunnelling currents the line of current drop may not coincide with the true fibre edge. When dealing with a step-like topography at the fibre edge, an additional lateral offset may occur because of the non-vanishing opening angle of the SFM tip. These involvements are depicted in Fig. 15. The zone between the true borderline and the line of current drop may be referred to as dead zone $\overline{DZ}$. For the sake of simplicity, the tilt of the tip occurring during scanning of the cantilever was neglected.

Given a perfectly flat surface and taking into account the very limited range of tunnelling currents, the upper bound of the width d of the dead zone $\overline{DZ}$ can be estimated to ~3 nm (Fig. 15b). Hence, the effective radius of the electrical tip-sample interaction area is larger than the geometric contact radius a. This is in analogy to the Maugis-Dugdale model (M-D) for purely mechanical interactions, where the surface forces are presumed to be limited to a ring $a<r<c$ around the contact area of radius a. When running experiments under ambient conditions, electrical currents might also flow along the water meniscus surrounding the close tip-sample contact. In the case of a step-like topography at the fibre edge (Fig. 15c), the total width d is given by the sum of both the reach of tunnelling currents and the geometric distance resulting from the step height and the opening angle of the tip.

When comparing with the location of the true fibre edge, the general conclusion is that the fibre-polymer borderline as defined by the current criterion exhibits a small lateral shift towards the polymeric material. Consequent-

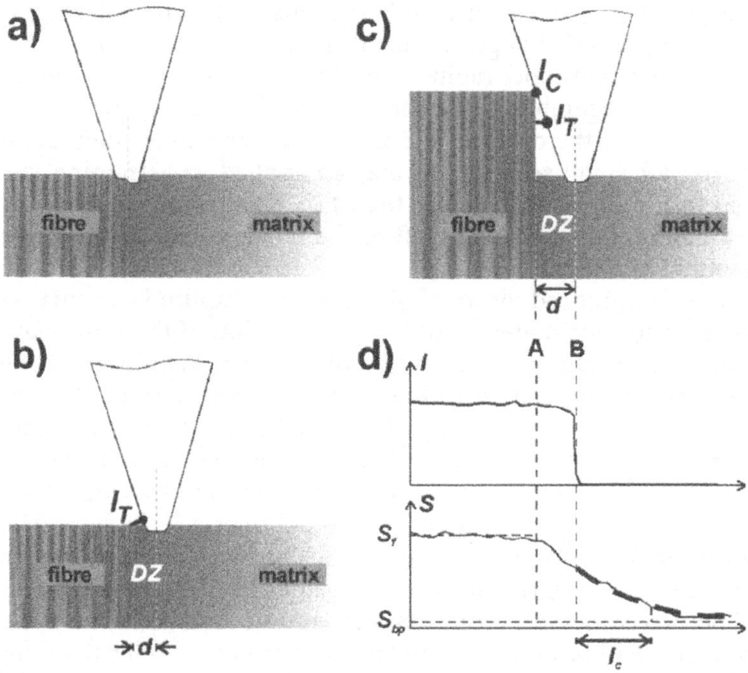

Fig. 15a–d Schematic representation of the tip-sample contact when the tip is close to the fibre edge [186]: **a** the tip is located just on the fibre edge. Owing to the strongly different mechanical properties of fibre and polymer, the deformation of the apex of the tip is expected to be asymmetric; **b** tunnelling currents I_T occurring between the fibre edge and the tip; **c** in the case of a topographic step mechanical contacts between the flank of the tip and the fibre may occur which have to be considered as the pathway for contact currents I_C; **d** analysis of the stiffness data was only performed within regions were the electrical current has collapsed

ly, stiffness variations occurring within the (unknown) dead zone $\overline{DZ}$ will not be taken into account when evaluating FMM amplitude profiles truncated according to the current criterion.

Whilst this fact imposes some limitations onto the minimal detectable width of the interphase stiffness gradient, it should be noted that the FMM measurement with the tip-sample contact area next to the fibre edge will in any case be affected by the presence of the nearby stiff fibrous material (Fig. 15a). This is due to the lateral extension of the sub-surface stress distribution which is roughly given by the double value of the tip-surface contact radius a [206]. Given the above value of 11 nm for a, the FMM amplitude gradient reflects solely PPS properties at a distance of ~22 nm away from the true fibre edge. As the stiffness gradients measured on PPS are much larger, they may not be ascribed to measurement artefacts due to the presence of the stiff fibrous material.

After having truncated the FMM amplitude gradients in accordance to the current criterion, they were fitted using an exponential profile:

$$S(N) = S_{bp} + S_w e^{-\frac{N-N_0}{l_c}} \tag{15}$$

The denotation is in close analogy to the semi-Gaussian stiffness profiles used for fitting the amplitude profiles measured on epoxies. In this case, N_0 corresponds to the position of breaking-down current amplitude and was set to zero. The mean value of l_c resulting from the fitting is 35.8±18.8 nm. Hence, the mean value of the total width of the interphase stiffness gradient is $3l_c \approx 107$ nm. Owing to the strong scatter in S_{bp}, it was treated as a free fitting parameter (which is in contradiction to the fitting procedure for the stiffness data measured on epoxy where S_{bp} was set to the mean bulk value). The pronounced variations of S_{bp} might originate from the partially crystalline nature of PPS, because regions of higher crystallinity are known to be packed more densely and to exhibit an increased stiffness. Some oscillatory stiffness profiles which were observed, supposedly reflecting both amorphous and crystalline regions, were not subjected to the fitting procedure. The oscillatory stiffness profiles may not be ascribed to instabilities of the topography feedback, since the variations of the topographic slope angle are rather low and inspection of the image of static force with respect to oscillations was negative.

4.5.3
Discussion of the Detected Stiffness Profiles

The SFM stiffness measurements on amine-cured diglycidylether of bisphenol-A (DGEBA)-based epoxies revealed that near the interface to Cu the epoxy stiffness is diminished as compared with the bulk epoxy stiffness. According to the rationale of interface-induced segregation of resin and curing agent superimposed on the chemical crosslinking reaction (diffusion-reaction model), the detected stiffness gradients result either from enrichment or depletion of the curing agent at the surface of the Cu phase. In order to shed some light on the correlation between epoxy stiffness and concentration of the curing agent, microscopic indentation experiments were conducted on a series of epoxy samples with differing concentrations of the curing agent [185]. Evaluation of the initial slopes of the unloading curves indicated an increase of stiffness of the cured epoxy Epikote828 (Shell) with increasing content of the curing agent L181 (Scheufler). The inflection point of the sigmoidal characteristic occurred at ~14.3 wt% of the curing agent.

As a consequence of these findings, in terms of the diffusion-reaction model the observed stiffness gradient indicates a depletion of the curing agent near the surface of the Cu phase. The significant variations of the characteristic length l_c ranging from ~100 nm to ~10 μm, may be traced back to different ratios of the characteristic times for reaction (t_r) and diffusion (t_d) or to different values of the diffusion constant D_T. Moreover, variations of the oxidation states of the Cu surface and of its morphology are expected to affect the driving force for diffusion. The temperature of the mixture of curing agent and epoxy resin was ~50 °C in each case, so that variations of D_T related to temperature seem less probable. Rather, variations of D_T due to different chemical compositions seem probable: whereas the epoxy matrix of the replica-sample (Fig. 10) was made up of the resin Epikote 828 and the curing agent EpikureF205 (both supplied by Shell), the epoxy matrix of the cross-sectional samples (Figs. 11, 12 and 13) were made up of the components L180 and H181 (both supplied by Scheufler).

As elucidated above, within the frame of the diffusion-reaction model the generation of an interphase requires comparable characteristic times t_d and t_r. A narrow interphase can be expected if $t_r<t_d$, for the increase of viscosity due to crosslinking occurs fast and diffusional flow may only occur along short distances. However, if $t_d<t_r$ the crosslinking-induced freezing of diffusional flow occurs later and wider interphases can be expected. Consequently, the width of the interphase depends on the ratio t_d/t_r. In addition to t_d, t_r may depend on time as well, because of the conversion of epoxide groups slows down after having started the curing reaction [207]. After having established some kind of concentration gradient, the ratio t_d/t_r may change along the gradient, as the reaction rate depends on the concentration of amines. It may be assumed that within the frame of superposed diffusion reaction processes, the kinetics plays a major role and the final width of the interphase depends strongly on the characteristic times for diffusion and reaction.

As suggested in Fig. 16, in the case of segregation of resin and curing agent both a region of enrichment and a region of depletion seem reasonable. However, all stiffness gradients detected are monotonic. Due to the dead zone next to the Cu/epoxy borderline where evaluation of the FMM data is not possible without ambiguity, the existence of a narrow zone next to the borderline with stiffness properties other than the detected ones may not be ruled out completely. In terms of the denotation of Fig. 16, in that case the detected stiffness gradients have to be attributed to the depletion zone and within the narrow enrichment zone the local stiffness is increased as compared with that of the bulk epoxy.

Contrary to the diffusion-reaction model, stiffness profiles resulting from temperature gradients established during the crosslinking reaction are expected to exhibit monotonic behaviour. As temperature variations affect the curing reaction and via this the final density of crosslinks, temperature gradients may result in respective stiffness gradients. During the exothermal curing reaction deviations of the local temperature from the mean sample temperature will occur in regions where the reaction heat is conducted away more efficiently than in other regions. This may exactly occur in regions close to the Cu phase (Fig. 17). Due to its heat capacity, reaction heat can be absorbed from the Cu particles. When dealing with continuous Cu components (e.g. in the form of a film) the reaction heat can even be conducted away efficiently, owing to the considerable thermal conductivity of Cu (bulk value 384 W/mK [208]).

The resulting decrease in temperature in regions close to the surface of Cu components may also affect segregation effects. In that case the effect of

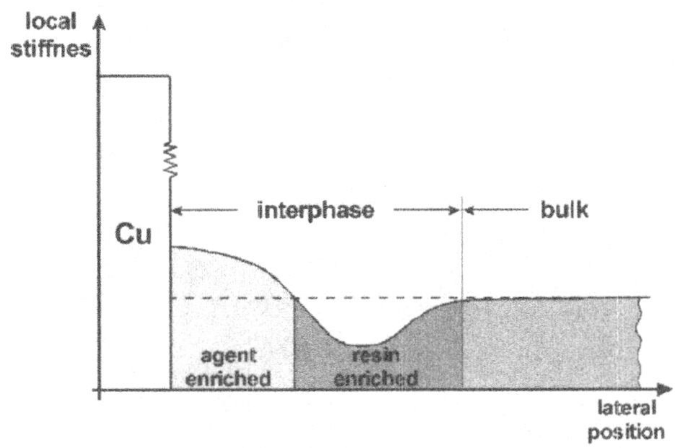

Fig. 16 Schematic representation of an interface-induced segregation scenario. As long as the viscosity of the crosslinking system is low enough, segregation of epoxy resin and curing agent may occur, driven by the polar surface of the Cu component. Conservation of mass requires a depletion zone close to the zone of enrichment. Via the network structure the concentration profile is reflected by the local mechanical properties of the cured epoxy system

the Cu components is twofold: they induce segregation effects (diffusion of molecules) as well as temperature gradients (diffusion of heat). The latter scales with the size of the Cu components, as the amount of heat absorbed and conducted away depends on the volume of the Cu particles. Thus for nanosize Cu particles the temperature effect is expected to be negligible.

Furthermore, the possibility of degradation effects should be taken into account. Oxidation reactions occurring at Cu surfaces are known to produce metal salts and complexes which are supposed to diffuse into the polymer matrix and may impair the interfacial strength [209]. Furthermore, it was suggested that Cu may induce the polymerisation of non-catalysed DGEBA [209]. Presumably, the oxides CuO and Cu_2O are of major importance for these effects.

As compared to the interphases detected in the case of the Cu/epoxy systems, the interphase measured in the case of the C-fibre/PPS system exhibits a negative stiffness gradient, i.e. decreasing local stiffness of the thermoplastic PPS with increasing distance from the C-fibre surface. The mean width $3l_c{\sim}107$ nm of the stiffness profile is ~2.6 times smaller than that of the stiffness profile measured on the Cu/epoxy replica sample. Taking into consideration the spatial constraints imposed on polymer chains due to the presence of the nearby hard wall represented by the surface of the C-fibres, the observed increase in local stiffness can be ascribed to the respective loss in chain flexibility and mobility.

For instance, this kind of quasi-immobilisation may result from multiple adsorption of one polymer chain at the fibre surface [210]. As shown by Bitsanis and Pan [211] by means of molecular dynamics (MD) simulations surface-induced ordering of chain segments can be expected which is re-

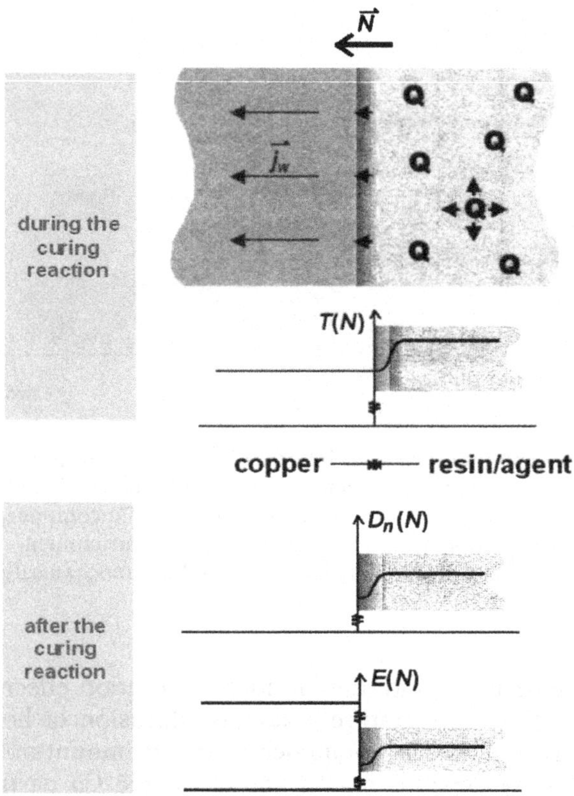

Fig. 17 Schematic representation of the temperature gradient occurring during the exothermic curing reaction. Owing to the comparatively high thermal conductivity of the continuous metallic filler (e.g. wires consisting of copper), exothermic reaction heat Q is conducted away from the interface. By changing locally the thermal conditions of the curing reaction, the temperature gradient $T(N)$ may influence the resulting network structure $D_n(N)$ which in turn defines the elastic properties of the epoxy, e.g. its elastic modulus $E(N)$. $\mathbf{j}_w$ denotes the heat current density

flected by the respective density profile exhibiting oscillations superimposed on a decay. The degree of ordering depends on the interaction energy with the hard wall surface. The density variations come along with different relaxation times indicating the restrictions imposed onto segmental mobility [211]. Dynamic-mechanical analysis (DMA) experiments performed at several polymer matrix composites by Tsagaropoulos and Eisenberg [212] delivered an additional glass transition attributable to chain segments with reduced mobility. The additional glass transition occurred at a temperature higher than that of bulk polymer. The second glass transition temperature was correlated positively with the mean molecular weight of the polymer.

Since the restriction of chain conformations is more pronounced for long polymer chains, some enrichment of short chains at the C-fibre surface seems probable. According to calculations made by Kosmas [213] the mean number of contacts between a surface and a nearby polymer chain is proportional to its molecular weight. As shown by Pangelinan et al. [214] the width of the interphase resulting from molecular weight segregation amounts to ~3.5 times the radius of gyration of the component exhibiting the highest molecular weight. Correspondingly, for typical thermoplastic polymers the interphase width is ~100 nm, which is consistent with the value deduced from the FMM measurements.

In the case of partially crystalline polymers such as PPS a higher degree of crystallinity may be expected close to the fibre surface. This is due to the fact that the necessary rearrangements of polymer chains for establishing crystalline order can be achieved more easily by short chains rather than long ones. A higher degree of crystallinity, however, corresponds to a closer packing of the polymer chains and, hence, to a higher local stiffness. Thus, the stiffness increase detected near the C-fibre can be attributed to both a reduction in chain flexibility and some chain ordering.

5 Mapping of Mechanical and Thermal Properties of a Model Brake Pad

5.1 Combined Measurement of Friction and Stiffness on a Model Brake Pad

With the aim of gathering detailed mechanical and thermal input data for micro- to mesoscopic modelling of frictional brake systems, characterisation of a brake pad with a strongly simplified formulation was performed by means of SFM. The study encompasses a thermal and various mechanical contrasts [215]. The manifold mechanical contrasts obtained with one experimental set-up allow some cross-checks and correlations to be performed among the available images, thus facilitating data interpretation.

The numerous components of friction pads for automotive brake systems can be classified roughly as abrasives, solid lubricants, reinforcing fibres, space fillers and the binder resin [216]. When designing the brake pad formulation, the types and relative amounts of ingredients have to be matched to several technical design criteria such as noise propensity, brake-induced vibration, reliable strength, wear resistance, as well as high and stable friction force [216]. The braking performance is governed by a complex interplay of the action of particulate additives, chemical reactions, thermal decomposition of organic ingredients and wear mechanisms. Using pad-on-disc-type tests, dynamometer tests or in-vehicle tests, both commercial brake pads and simplified formulations were studied. Besides the macroscopic simulation of the operational conditions of technical systems, a more fundamental approach can be devised. From the microscopic point of view, the surface of the rotating grey-cast iron disc of a disc brake can be modelled as an ensemble of asperities covering a broad range of surface energies, stiffness values, heights and widths. When in action, the brake pad is rubbed against the rotating disc. Instead, by scanning a single asperity over the surface of a brake pad of well-known composition, one may get some basic information concerning the mechanical and thermal interactions between the asperity and the particulate constituents of the brake pad. The resulting data may allow a more detailed description of the microscopic parameters governing frictional processes and may contribute to the formulation of elaborate micro- or mesoscopic models with a rather low degree of mathematical homogenisation. Recent efforts for developing simulation techniques based on mesoscopic particle methods [217] seem to be a promising approach for dealing with material inhomogeneities.

Having prepared a highly simplified system consisting of a phenolic resin matrix, antimony sulfide (Sb_2S_3) particles and polybutadiene (PBD) rubber globules, topography, lateral force, modulated lateral force, stiffness and thermal diffusivity of the sample surface were imaged. Owing to the simple formulation of the sample and to the pronounced differences in particle size and shapes, the domains prominent in the SFM images can be assigned easily to the two different materials (Sb_2S_3 and PBD) dispersed in the phenolic matrix. The three ingredients used for producing the simplified brake pads were supplied by a manufacturer of automotive brake pads (AlliedSignal Bremsbelag GmbH, Glinde, Germany). The typical diameter of the rubber globules was $\sim$50 μm. Details of sample preparation are described in [215].

In Fig. 18 SFM images are given as measured on a polished cross-sectional surface of the model brake pad. The figure contains the images of topography (Fig. 18a), static normal force (Fig. 18b), static lateral force as measured in both scanning directions (Fig. 18c,d), amplitude and phase of the dynamic lateral force (Fig. 18e,f), as well as amplitude and phase of the dynamic normal force (Fig. 18g,h). The scanned area is 80×80 μm^2.

The circular-shaped domains can be attributed to the globular PBD particles. Hence, the much more irregular shaped and smaller islands can be identified as the cross-sections of the Sb_2S_3 particles. For reasons of comparison, the mean values of the signal magnitudes measured on each of the constituents are displayed in histograms where the offset suppressed in the images was taken into account.

In Fig. 18c,d a significant contrast between PBD and the phenolic matrix is observable. Obviously, this contrast is inverted when the scanning direction is reverted. This is due to the fact that the sign of the angle of cantilever torsion changes with the direction of scanning (friction loop). By virtue of negation of the image measured in reverted direction, the contrasts due to different materials are consistent whereas the contrasts due to topographic features are inverted. In comparison with the images of the static lateral force, the amplitude image of the dynamic lateral force (Fig. 18e) clearly exhibits small irregular islands attributable to the Sb_2S_3 particles. The M-LFM phase image (Fig. 18f) is dominated by the contrast between the PBD globules and the phenolic matrix. This is similar to the phase image of the FMM measurement (Fig. 18h). However, on Sb_2S_3 the FMM phase signal seems to be less deteriorated by topographic features. On larger grains of Sb_2S_3 the FMM phase shift is discernibly lower than on the phenolic matrix.

From the pronounced phase shift measured on PBD, it may be inferred that the time scale of both the lateral and the normal displacement modula-

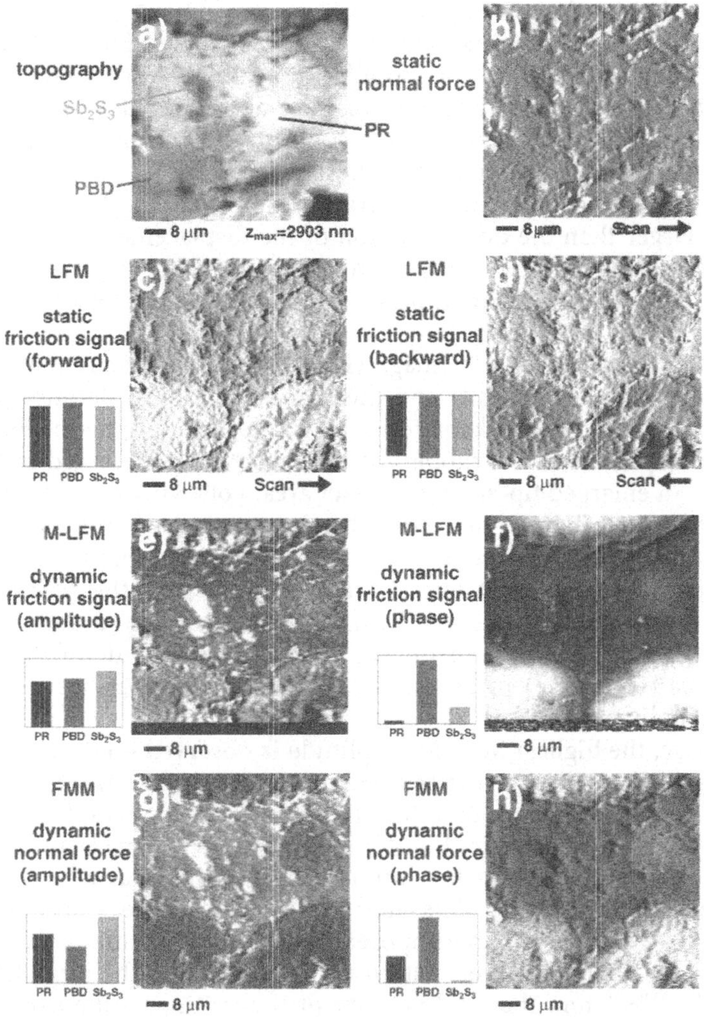

Fig. 18a–h Combined M-LFM and FMM measurements performed on a model brake pad sample [215]; **a** topography; **b** static normal force; **c,d** static lateral force signals; **e,f** signals of the M-LFM; **g,h** signals of the FMM, as measured on a sample consisting of polybutadiene (PBD), antimony sulfide (Sb_2S_3) and phenolic resin (PR). High signal values correspond to light grey values. Scan size 80 μm, scan velocity 50 μm/s. The displacement modulation measurements were performed at 61.1 kHz (M-LFM) and at 59.0 kHz (FMM), respectively

tion measurements (cycle duration $\sim$17 μs) lies within the glass-rubber transition regime of PBD. Owing to the kinetic nature of this transition, the transition regime encompasses several orders of magnitude of relaxation times. According to the time-temperature superposition principle, the glass-rubber transition can be sampled by adapting the loading rate to the relaxation times typical for the temperature of the experiment. As a result of the comparatively low glass transition temperature of PBD (T_g~−106 °C [218]), at room temperature the characteristic times of molecular rearrangements are much shorter than 1 s. In contradiction to PBD, the temperature of the glass-rubber transition of phenolic resin is above room temperature. Depending on the curing scheme and the concentration of the curing agent, the onset temperature of the transition lies between 60 °C and 150 °C [219]. Consequently, the characteristic relaxation times of the phenolic resin are distinctly larger than the cycle duration of the displacement modulation experiments and the material is expected to behave in a glassy manner. This is in agreement with the observation of a total phase shift lower than that of PBD.

From the FMM amplitude image (Fig. 18g) it is obvious that the local stiffness is the lowest on PBD and the highest on Sb_2S_3. Because the static part of the normal load is kept constant via the topography feedback loop, the high compliance of the rubbery PBD results in an increased indentation depth and an enlarged tip-sample contact area. Following the adhesion model of Bowden and Tabor [150, 220], this in turn causes an enlarged lateral force. Within this model, the friction force is proportional to both the real area of contact and a mean lateral force per unit area, the so-called shear strength (Sect. 3.1.4). Consistently with the conclusion drawn from that model, both the magnitude of the static and the amplitude of the dynamic lateral force (Fig. 18c,e) measured on PBD are larger than those measured on the phenolic matrix.

Otherwise, the highest M-LFM amplitude is observed on the stiff particles consisting of Sb_2S_3. Due to the corrugation of the surface and to the susceptibility of the static lateral force for cross-talk from topography, the magnitude of friction force measured on Sb_2S_3 remains rather unclear. The M-LFM signals, however, are much less prone to topographic artefacts and show clear contrasts [221, 222]. This effect is caused by the sinusoidal lateral motion of the cantilever superimposed on the linear scanning motion. Due to the chosen frequency of excitation (f=61.1 kHz) and the modulation amplitude of x^{dyn}~1 nm, the total velocity of the motion of the cantilever has both positive and negative values. Thus, during each modulation period the tip moves both forward and backwards and goes through a complete friction loop. In practice, the complete elimination of the cross-talk from topography may be achieved only if the respective lateral forces are independent of velocity and if the cantilever, especially the position of its tip, is perfectly symmetric to its length axis.

In the case of the comparatively stiff Sb_2S_3, the high friction level cannot be ascribed to the large contact area. In terms of the adhesion model, the high amplitude of the dynamic lateral force results from a high shear strength. This conclusion seems to contradict the low friction behaviour that one might expect of solid lubricants. However, as shown by Singer [223], the effectiveness of solid lubricants is determined by interfacial films rather than the bulk shear properties of the lubricant itself. The interfacial films can be generated during sliding. Thus, it may be inferred that during the mechanical polishing procedure and the SFM experiments no friction film has been generated.

Finally, it should be noted that via increased frequencies, high shear velocities can be implemented in the nanoscopic M-LFM experiment. When applying excitation frequencies in the MHz range, the amplitude of shear velocities approaches values of the order of 0.01 m/s, even for small displacement modulation amplitudes of the order of some nanometres.

5.2 Thermal Diffusivity Measurements on a Model Brake Pad

The mesoscale temperature distribution occurring in a composite surface during frictional loading depends on a set of mechanical and thermal parameters that have to be accessible when a well-directed optimisation of the formulation of brake pads is intended. After all, the performance and the wear behaviour of such systems are strongly dependent on the temperature distribution occurring during operation. Hence, in addition to the mechanical SFM measurements, scanning thermal microscopy (SThM) experiments were performed with the aim of mapping the thermal diffusivity. Along with topography, the images of thermal diffusivity are given in Fig. 19. With a certain amount of overlapping, the sample position was translated after each measurement. By composing the resulting set of scans, a more extended area of the sample surface is visible (stitching). The scan range of each image was 80 μm. Owing to the comparatively large radius of curvature of the tip designed for thermal measurements, the lateral resolution of the topographic image at Fig. 19a is lower than that of that at Fig. 18a.

From the topographic composite image at Fig. 19a some hole-like depressions are recognisable. Most of them can be ascribed to the pores of the phenolic matrix. Similar to the mechanical contrasts, the image of thermal diffusivities is disturbed in the region of severe depressions. The significant increase in apparent thermal diffusivity indicates an enlarged probe-sample contact area. In particular, this effect can be expected when the arms of the V-shaped Pt/Rh filament representing the tip of the SThM probe touch the side walls of the hole. From the cross-sections shown in Fig. 19 it is obvious that the width of the respective hole is comparable to the radius of curvature of the thermal probe. The contour of the probe was deduced from an SEM micrograph. In contrast to concave holes, regions with a convex curvature exhibit a reduced apparent thermal diffusivity.

Neglecting regions with severe topographic features, the thermal diffusivity is the highest on PBD and the lowest on the phenolic matrix. Compared with the latter, the areas consisting of Sb_2S_3 also exhibit an increased diffusivity. The ratios of the macroscopic thermal diffusivities $\lambda_{th}/c_{th}\rho$ of phenolic resin, Sb_2S_3 and PBD are 1.0:1.7:5.3 (thermal conductivities λ_{th}~0.27, 0.25 and 0.22 W/mK [224–226]; c_{th} is the specific heat capacity; ρ is the mass density). Thus, the detected microscopic diffusivities are qualitatively consistent with data originating from macroscopic measurements. Quantitative deviations might be caused by the magnitude of the effective tip-sample

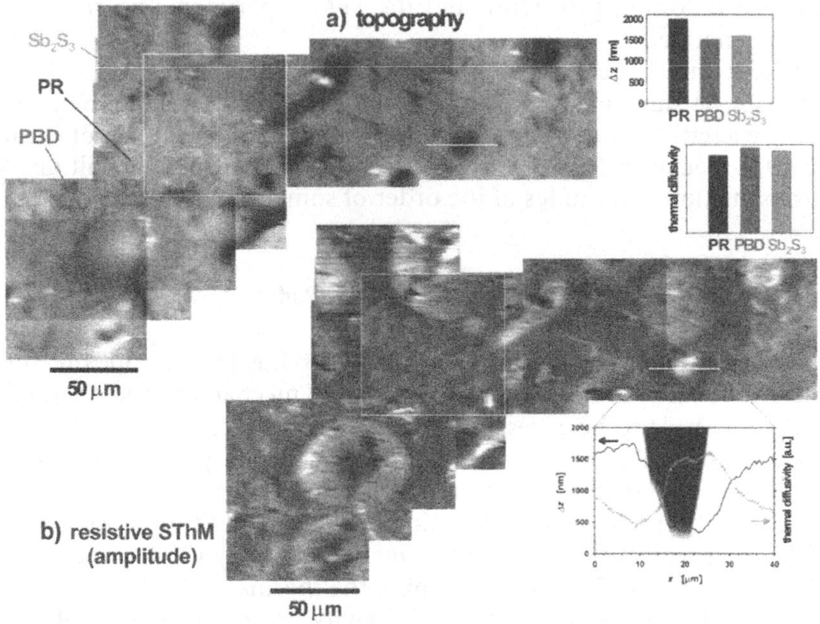

Fig. 19 a Topography. b Thermal amplitude signal related to thermal diffusivity, as measured on a model brake pad sample consisting of polybutadiene (PBD), antimony sulfide (Sb_2S_3) and phenolic resin (PR) [215]. High signal values correspond to *light grey values*. The scanning area was shifted successively. The averaged data displayed in the histograms result from evaluating the image marked with a *white borderline*. The *white horizontal bars* mark the location of the cross-sections shown in the graph. This also includes the appropriately scaled contour of the thermal probe as deduced from an SEM micrograph. The radius of curvature of the probe is similar to the lateral extension of the hole, approximately. Scan size of single scans: 80 μm, scan velocity 40 μm/s. The modulation of the heating current passing the Pt/Rh filament was performed at 132 Hz

contact area, or by the thin water layer covering the surface of both tip and sample.

In principle, the heat conduction can occur via near-field radiation, the gas molecules of the surrounding atmosphere, the liquid film covering tip and sample surfaces and forming a liquid bridge, or the mechanical tip-sample contact. By means of surface roughness, the effective contact area is presumed to be much smaller than the apparent (i.e. the geometric) one (Fig. 20).

The effective contact area rises with increasing compliance and plastic deformability of the contacting materials. In the case of macroscopic polymer-aluminium junctions, the increase of thermal diffusivity with rising pressure was demonstrated by Marotta and Fletcher [227]. According to Parihar and Wright [228], junctions adjoining elastomeric materials are already free of cavities for pressures as low as 0.05 MPa. Bearing in mind the low stiffness of PBD, a close tip-sample contact and consequently a good thermal conductance have to be expected.

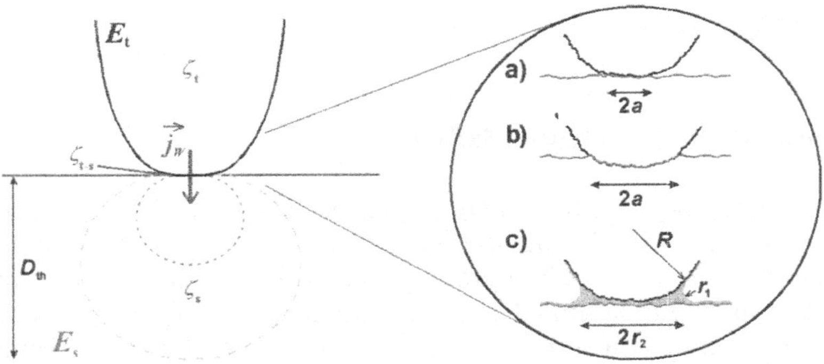

Fig. 20a–c Schematic representation of the thermal conductances of tip (ζ_t), tip-sample-contact (ζ_{ts}) and sample material (ζ_s) [215]. Along with the frequency of the heating current, the penetration depth D_{th} of the thermal wave depends on the thermal resistance of the configuration. E_t and E_s denote the Young's modulus of the tip and the sample, respectively. The enlarged contact situations represent: **a** a dry and stiff; **b** a dry and compliant; **c** a moist and stiff contact. a is the mechanical contact radius and r_2 is the radius of the liquid bridge. R and r_1 denote the radius of curvature of the tip and of the liquid meniscus, respectively

As argued by Luo et al. [229], under ambient conditions the dominant tip-surface heat conduction is most likely through a liquid film bridging tip and sample. Hence, the spatial resolution of thermal measurements is limited by the sharpness of the thermal probe and the thickness of the water film on the surfaces of tip and sample. Since the lateral extension of the water film surrounding the tip-sample contact (as given by $2r_2$; see Fig. 20) increases with decreasing contact angle, the thermal conductance of the tip-sample junction is expected to rise with the hydrophilicity of the surface under investigation. With increasing surface roughness, the effective contact area diminishes and the importance of water condensed in the capillaries is supposed to increase. When operating under ambient conditions, in general the contacting surfaces are separated by a few monolayers of water [230]. Owing to the layered structure of the mediating water film, its thermal conductivity may deviate from that of bulk water [229]. Finally, due to the lubricating behaviour of water layers the friction measurements are affected by condensed water as well [231].

Under conditions of strong frictional loading, the interplay of mechanical and thermal properties may play an important role. Via positive thermal expansion, frictional heating leads to increasing loading of the rubbing surfaces which in turn increases the frictional forces and the respective heating. This kind of intrinsic amplification is commonly referred to as frictionally excited thermoelastic instability (TEI) [232] and is considered as one of the possible causes for the frequently observed oscillations of braking forces [233, 234].

6 Nanolithography

6.1 Mechanical Lithography on Polymer Surfaces

Using lithography in contact mode the surface is modified by increasing the applied normal load during the scan. The tip indents the sample and scratches the surface. This lithographic method is called also *static plowing*.

Contact mode lithography has been applied most of all to hard materials: mica [235], semiconductors [236, 237] and metals [238–240]. Langmuir-Blodgett films [241, 242] and thin polymer films [243–251] can also be structured in contact mode.

Performing contact mode lithography on polymer films presents several drawbacks. First of all, only thin films (5–25 nm) on hard substrates can be modified. In addition, the scanning of the modified surface can be performed only with small forces, and hence with a reduced spatial resolution. If the force is too large (the threshold depends on the polymer and on its thickness), the polymer will be further modified. The only solution is to scan the surface after the modification in another mode. However, there is another serious drawback: the torsion of the cantilever during the lithography process generates irregularities in the profile of the edges. For this reason the direction of the lithographed lines is limited to a certain range around the axis of the cantilever [252, 253].

Due to the numerous problems of contact mode lithography, several researchers have tried to modify this technique and to perform lithography by modulating the force between tip and sample. The first lithography experiment pioneering this technique is that of Jung et al. [253]. The authors make a comparison of three different techniques (force-displacement curves, contact mode and a dynamic technique). The authors claim that a modulation of the contact force is necessary in order to eliminate the irregularities in the topography of the lithographed structures that are caused by the torsion of the cantilever and by *stick-slip*. Such a modulation is obtained by scanning the sample vertically with a frequency of 4 kHz. This method is very different from a real tapping mode lithography. As a matter of fact the resonance frequency of the cantilever is normally very much larger, and the transfer of oscillations to the cantilever depends on the stiffness of the sample. For this reason the energy transfer is generally not very efficient, and differences in the modification process cannot be excluded.

Magno and Bennett [236] have modified the contact mode lithography by giving an audio frequency signal to the *dither-piezo* only when the sample has to be modified. The advantage of this method is that the topography can be acquired or modified without changing the scanning mode.

Another group of experiments [254–256] was intended to perform lithography in tapping mode. Even if these techniques are very similar to Dynamic Plowing Lithography, they are, strictly speaking, not exactly the same, and will be reviewed here. The increase of the force between tip and sample is obtained by giving an additional signal to the *dither-piezo*. This signal makes the tip approach to the sample. The authors point out once more that, thanks to this technique, the irregularities in the topography of the lithographed structure can be eliminated.

6.2 Dynamic Plowing Lithography

6.2.1 *State of the Art*

Dynamic plowing lithography (DPL), i.e. the lithography technique in tapping mode in which the force between tip and sample is increased by suddenly increasing the amplitude of the cantilever oscillations, has been developed by Klehn and coworkers [257–260]. The topography of the sample is acquired with a normal, small amplitude of the cantilever oscillations. When the amplitude is increased, the feedback makes the sample approach to the tip, in order to keep constant the oscillation amplitude, and the tip indents the sample.

The transfer of a certain pattern on the surface can be performed in *vector mode* or in *scanning mode*. In vector mode, the software provides a set of commands that permits one to write lines of arbitrary length and direction with defined scan speed and oscillation amplitude. In this mode the whole surface is previously scanned without performing lithography, i.e. with small oscillation amplitude, in order to acquire the topography. In vector mode the hysteresis of the piezoactuators has an important effect, and the pattern can be transferred exactly only with *closed-loop* microscopes. The scanning mode is a synchronisation of the raster scan mode with the desired pattern.

DPL presents the following advantages:

1. In DPL, thanks to tapping mode, it is still possible to image and structure soft samples with relatively low forces using cantilevers with higher spring constants.
2. In DPL, it is possible to eliminate the irregularities of the topography of the modified structures that are one of the major problems of contact mode lithography.
3. In DPL, it is possible to indent a surface and immediately image the indentation, taking advantage of the backward scanning image. This in-situ imaging ability eliminates the need to move the sample, to change tips, to relocate the area for scanning or to use a completely different instrument to image the indentation.

6.2.2
Experimental

A schematic representation of the experimental set-up necessary to perform DPL is given in Fig. 21. DPL is a lithography technique, in which the microscope is operated in tapping mode (see Sect. 2). When the microscope is operated in tapping mode, the cantilever driven by a dither piezo vibrates near its resonance frequency. In conventional tapping mode the amplitude of the oscillations is chosen so that the tip of the cantilever taps intermittently the sample surface. In dynamic plowing lithography, modifications of the sample are obtained by increasing the modulation amplitude applied to the dither piezo that drives the cantilever oscillations. By increasing the oscillation amplitude the sample is moved nearer to the tip and the tip is likely to indent the sample surface, producing elastic and plastic deformations. The oscillation amplitude is changed by means of the function generator that controls the voltage given to the dither piezo.

The oscillation amplitude values are encoded in a pixel image called the mask. The signal used to write, V_w, is roughly a 20-fold factor larger than the signal used to read, V_r. Thus, when the cantilever is far from the surface, the writing free oscillation amplitude, A_{wf}, is 20-fold larger than the reading free oscillation amplitude, A_{rf}. The efficiency of the writing process depends on the difference $A_w - A_r$, where A_r and A_w are the reading and writing amplitudes when the tip is close to the surface. This difference depends on $A_{wf} - A_{rf}$, but also on the damping of the cantilever oscillations due to the interactions between the cantilever and the sample. The damping depends in turn on the set-point amplitude.

6.2.3
Results

Figure 22a shows a hole in polymethylmethacrylate (PMMA) written with a mask containing an isolated white pixel [261]. The hole is surrounded by a border wall of variable dimensions, higher than the unmodified surface. Both the shape of the hole and of the border wall depend on the tip shape. Also the position of the border wall with respect to the hole depends on the tip, but not on the scanning direction. The hole has the shape of the tip, i.e. a pyramid with an equilateral triangle as the base.

Figure 22b shows the dependence of the depth D_{DPL} and of the width W of the hole (height of the equilateral basis) on the voltage ratio V_w/V_r. The depth and the width of the holes are proportional to V_w/V_r, as expected.

Figure 23 shows the dependence of the volumes on V_w/V_r. Two volumes have been measured: the volume of the holes, i.e. V_{neg}, and the volume of the border walls, i.e. V_{pos}. The volumes follow a cubic law. The volume of a pyramid with an equilateral triangle as basis is given by $V = \frac{1}{4\sqrt{3}} S_H^2 D_{DPL} \cong 0.14 S_H^2 D_{DPL}$, where S_H is the length of the side. Using the measured experimental quantities S_H, D_{DPL} and V, the calculated proportion-

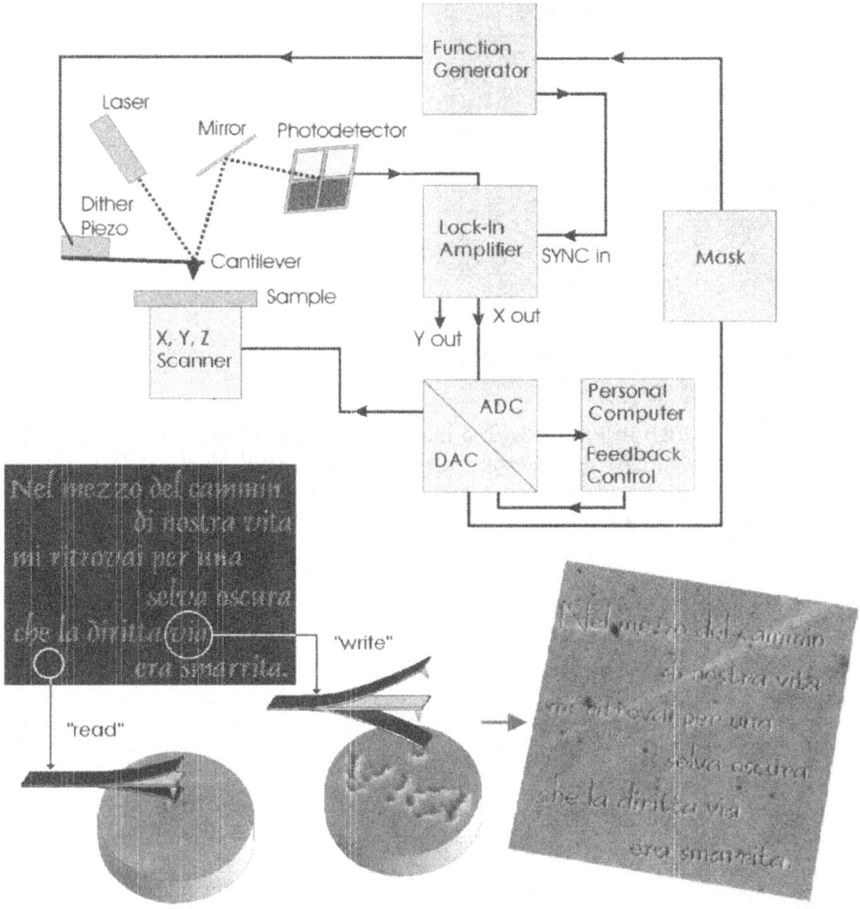

Fig. 21 Schematic representation of the set-up for DPL in *scanning mode*. The values of the amplitude of the cantilever oscillations are encoded in the mask and given to the dither piezo through the function generator. An example of mask, containing the first tercet of the Divine Comedy of Dante Alighieri (*Midway upon the journey of our life / I found myself within a forest dark, / for the straightforward pathway had been lost.* Translation of H. W. Longfellow), is shown in the *lowest part of the figure*: *white pixels* correspond to large oscillation amplitudes, causing the tip to indent the sample, and *black pixels* correspond to small oscillation amplitudes. On the *right*, the result of the sample modification (scan width 5 μm) is shown

ality factor between V and $S_H^2 D_{DPL}$ is ρ=0.16±0.02. This confirms that the convolution with the tip shape does not alter the shape of the holes, even if it may engender a certain error in the determination of the dimensions.

For V_w/V_r<10 no modification is obtained. For V_w/V_r=25, V_{neg} is approximately equal to V_{pos}. For higher modulation amplitudes V_{pos} is very much larger than V_{neg}. The elasto-plastic response of the sample can be described by means of the parameter

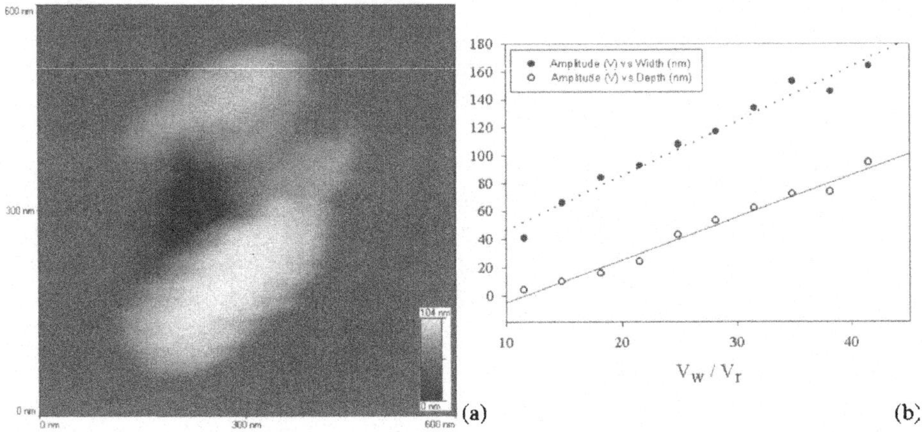

Fig. 22 **a** A hole written in PMMA with a mask containing an isolated white pixel (scan width 600 nm, *z* scale 104 nm, 60×60 pixels). **b** Width (*full black circles*) and depth (*open circles*) of the carved holes as a function of the writing modulation amplitude V_w/V_r. Both dimensions are proportional to the modulation amplitude

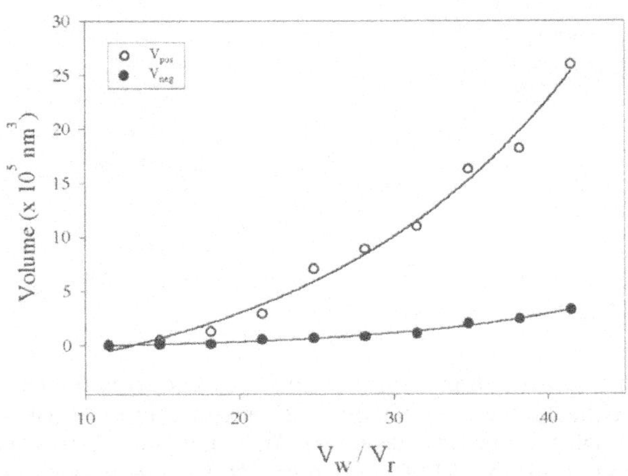

Fig. 23 Dependence of the volume of the holes V_{neg} (*full circles*) and of the border walls V_{pos} (*open circles*) on V_w/V_r. Both volumes follow a cubic law. For $V_w/V_r<10$ no modification is obtained. For $V_w/V_r \leq 25$ V_{neg} is approximately equal to V_{pos}. For higher modulation amplitudes V_{pos} is very much larger than V_{neg}

$$\psi_V = 1 - \frac{V_{pos}}{V_{neg}} \tag{16}$$

In conventional indentation and cutting processes $0<\psi_V<1$, with $\psi_V=1$ ($V_{pos}=0$) for ideal *microcutting* and $\psi_V=0$ ($V_{pos}=V_{neg}$) for ideal *microplowing*. DPL is characterized by $-6<\psi_V<-2$ for $V_w/V_r<25$ and by $-10<\psi_V<-6$ for $V_w/V_r \geq 25$.

As already stated, the convolution with the tip shape may cause a false determination of the volumes. However, the tip topography can be corrected through a *blind reconstruction* [262, 263]. The corrected images give larger values for V_{neg} and smaller values for V_{pos}, but, for $V_w/V_r \geq 25$, $\psi_V<0$.

In other words, depending on the scanning parameters, but above all depending on the given writing voltage V_w, the total volume can be increased. Reasons for an increase of the total volume will be discussed in Sect. 6.4.

Other parameters can be varied, for instance the number of contacts N between the tip and sample surface during the lithography. In order to tune this parameter holes have been written with different scan speeds at different writing voltages [261]. The linear dimensions of the holes increase linearly with the number of contacts, until they reach a plateau, whose height depends on the writing voltage V_w. The hole is carved during the first 40 or 50 contacts; during the successive contacts, the polymer is further compressed, so that the depth increases, but the width shows only little increase. When the voltage V_w is large enough, holes can also be obtained with only seven contacts. Also the border wall of the holes is already created at the beginning of the lithography process, during the very first contacts. For $N<20$, V_{neg} and V_{pos} are quite the same but, as the number of contacts is increased, the positive volume becomes much bigger than the negative volume.

The shape of the lines written with DPL can be observed in Fig. 24: the tip embosses a groove surrounded by border walls. The depth of the lines and the height of the border walls are in agreement with the experiments with single holes. Depending on the tip, it is possible to write 3 nm deep and 30 nm wide lines [264].

Lines can be written only perpendicular to the fast scan direction. Also the carving of large uniform surfaces, e.g. large stripes or squares, presents some problems [264]. Instead of carving, e.g. a stripe, the tip cuts a groove at the beginning of the stripe and a border wall is present at the end of the stripe. This border wall is the border wall of the line written at the beginning, but carried by the tip to the end of the stripe. In other words, the tip carves the surface only at the beginning of the region that should be written. This is due to the interplay between lithography and feedback. The increase of the oscillation amplitude can be considered as a disturbance for the feedback. The feedback has a finite reaction delay. Due to this delay, at the beginning the feedback does not change the position of the sample. After this, the feedback can slowly adjust the tip-sample distance in order to decrease the cantilever oscillations. Hence, deep indentation occurs only at the very beginning [264]. The fact that the feedback is antagonistic to lithography explains not only why stripes or squares are written only at the beginning, but also why horizontal lines cannot be written. An horizontal line is a sequence of "disturbances" along the fast scan direction, while a vertical line consists of only one "disturbance" for each scan line. Only in the first case does the feedback "have the time" to react to the increase of the oscillations.

Since uniform surfaces cannot be written by transferring a large uniformly white surface in the mask, such surfaces must be written without feedback. This presents some problems anyway because of the possible irregular-

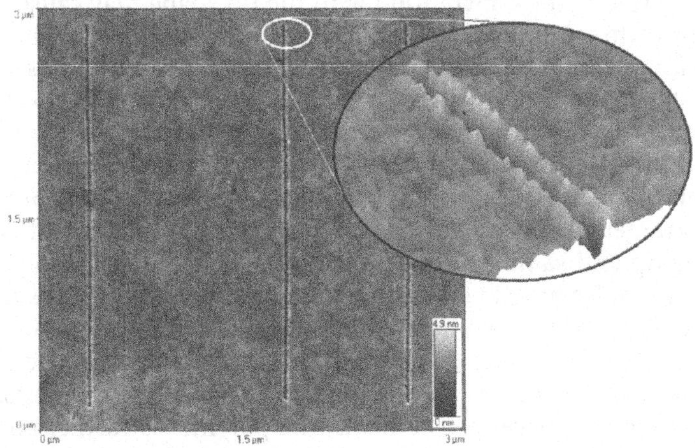

Fig. 24 Three lines written in PMMA with V_w/V_r=10. The lines are 40±3 nm, 38±4 nm and 37±2 nm wide, and 3.4±0.4 nm, 3.4±0.3 nm and 3.4±0.3 nm deep (scan width 3 µm, z scale 5 nm, 500×500 pixels). The *inset* shows a zoom at the beginning of the middle line

ities in the topography of the film. Another way to circumvent this problem is to carve a surface by writing sequences of overlapping holes (pulse train) [261]. Figure 25 shows two squares written with overlapping holes and different density of the holes, i.e. different distance between the holes. Each hole is the result of more than 200 contacts. The insets show the Z profile along the white lines. In the first square the holes overlap only partially. It is evident that only the first (left) vertical line and the last horizontal line present deep holes. This can be explained as follows. When the first horizontal line is written, the tip carves holes and a border wall is created below the hole and at its right side. When the second horizontal line is written, the holes overlap with the border wall of the already carved holes. The polymer molecules in the border walls of the holes are partly compressed and partly shifted to the right and the reached depth of the holes is smaller. Furthermore, depending on their position, the new border walls may partially fill the already carved holes. Since the border wall is created only below each hole and at its right, only the holes in the first vertical line on the left-hand side can reach the full depth, and these holes are not filled again by the border walls of new holes. Also the holes in the last horizontal line are not filled again. Thus, the first vertical line on the left-hand side and the last horizontal line result in more depth. At the right of the square a high border wall is created, because a part of the border walls is shifted to the right. When the squares are written with higher density, they present the same features: 1) the left and the bottom parts of the square are deeper; 2) the border walls of the single holes are accumulated at the right and below the square; 3) not all the material of the border walls of the single holes can be shifted to the right or below, and some material stays inside the square. The resulting structure is a rather uniform depression, with the left vertical side and the bottom

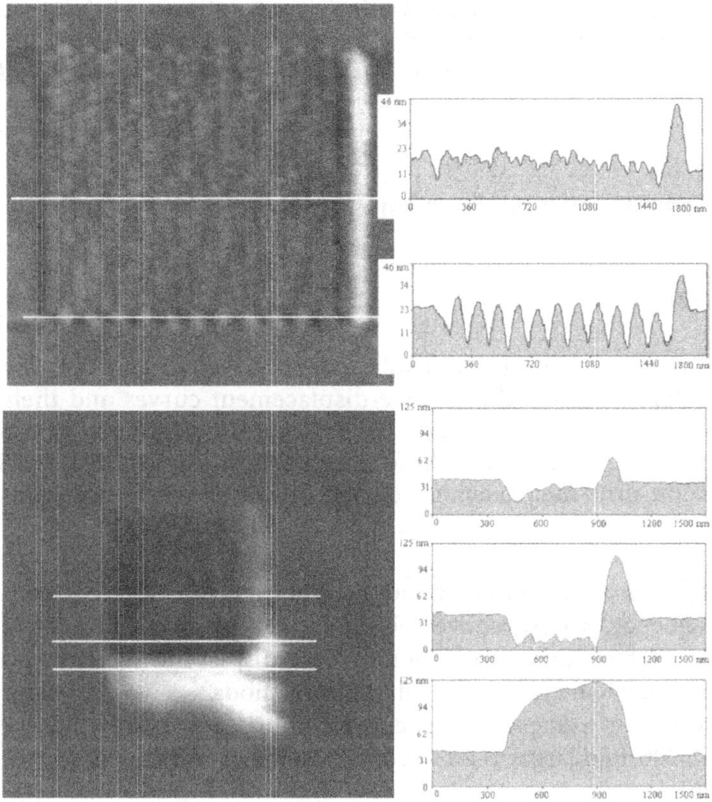

Fig. 25 Two squares written with masks containing alternating white and black pixels (pulse train) of different density. The *first square* has been written with a distance between two subsequent white pixels of 62 nm. The surface cannot be carved uniformly, as it can be seen in the height profiles on the right. Only the bottom horizontal side presents deep holes. The *second square* (distance between two subsequent white pixels of 19 nm) is more uniformly carved, even if the left vertical side and the bottom horizontal side are deeper. In the second case, a large and tall border wall on the right and above all under the square can be observed. This border wall is made up of the border walls of the single holes, that have been partially carried out by the tip. Some residues stay inside the squares, and only the deeper regions are free from residues. The three height profiles are taken inside the square, in the deeper bottom horizontal side, and on the border wall. For the first image the scan width is 1.7 μm, the *z* scale is 33 nm, and the resolution 500×500 pixels, for the second image the scan width is 1.3 μm, the *z* scale is 109 nm, and the resolution 80×80 pixels

horizontal side deeper than the rest, and a large and tall border wall at the right and below the square.

DPL has been proved to be an efficient tool to modify and structure polymer surfaces, permitting one to circumvent the numerous drawbacks of contact mode lithography. The antagonism between writing process and feedback can be eliminated by carving surfaces through overlapping holes (pulse-train). On the other hand, DPL leads to the formation of undesirable very high and large border walls. The structure and the properties of these border walls will be analysed after comparing DPL with another lithographic technique, namely force-displacement curves indentation (FDI).

6.3 Force-Displacement Curves and Force-Displacement Curves Indentation (FDI)

6.3.1 *State of the Art*

A short introduction on force-displacement curves can be found in Sect. 2. A more detailed review about force-displacement curves and their application is given in [6]. In this section we will focus our attention on the use of force-displacement curves as a lithography tool in comparison to DPL.

The use of force-displacement curves as a lithography tool presents two important drawbacks:

1. Since force distance curves cannot be acquired with any frequency (the normal frequency is 1 Hz; above 3–4 Hz oscillations may engender several important artefacts), the acquisition of force-displacement curves is very time consuming, when compared with other methods.
2. Since each force-displacement curve comprises normally 200 points, data files are 200-fold larger than in contact mode or in DPL.

Since the first experiment, where Bhushan and Koinkar [265] have measured the hardness of silicon with a modified SFM, researchers have been using force-displacement curves more for the determination of certain sample properties, e.g. hardness and stiffness, than for the structuring of surfaces.

Only in the last three years have some researchers begun to pay their attention to the structured surface and to parameters influencing the lithography with force-displacement curves (FDI) [266–268]. The big advantage of FDI is the possibility of gaining knowledge about the whole indentation process: during FDI, the force and the indentation are known at every point, and not only stiffness and hardness, but also other important properties such as density, elasto-plastic behaviour, adhesion, time behaviour, etc. can be measured and calculated.

6.3.2 *Results*

Our method for the acquisition of force-displacement curves is described in details in [32]. Force-displacement curves are acquired following the surface profile and the sample is indented until a maximum force F_{max} is reached.

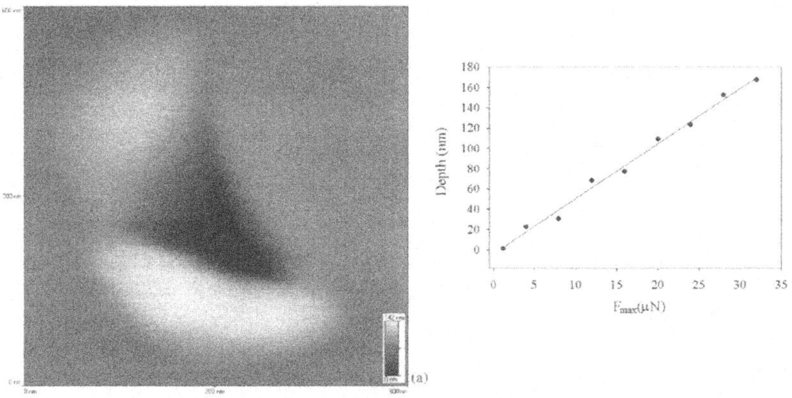

Fig. 26 **a** A hole written in PMMA with a single force-displacement curve (scan width 600 nm, *z* scale 142 nm, 60×60 pixels). **b** Depth of the carved holes as a function of the maximum force F_{max}. The depth is proportional to the maximum force

Figure 26 shows a hole carved with FDI (a) and the dependence of the depth of the holes D_{FDI} on the maximum force F_{max} (b). Holes written in FDI are a replica of the form of the tip. Also in this case the holes are surrounded by a border wall that is smaller than the border walls of the holes written in DPL.

Since the force-displacement curve contains information about the whole indentation process, the elastic deformation of the sample can be measured and used to calculate the stiffness $S=\mathrm{d}F/\mathrm{d}h$ at $h=h_{max}$, where F is the force and h is the indentation. As already explained in Sect. 3.1.1., in order to relate the stiffness to the Young's modulus, it is necessary to make assumptions about the contact area. The depth of the permanent indentation (plastic deformation), i.e. the depth D_{FDI} shown in Fig. 26b, and the maximum indentation (sum of the plastic and of the elastic deformation) can be used to calculate a parameter that describes the relative weight of the elastic and of the plastic response.

The dependence of the D_{FDI} on the number of cycles, i.e. on the number of indentations repeatedly performed on the same point (Fig. 27), is in agreement with the results obtained for DPL [263]. As in DPL, the depth reaches a plateau after 40–50 cycles, and the plateau depends on F_{max}.

Since both D_{FDI} and D_{DPL} depend linearly on V_w/V_r and on F_{max} respectively, it is possible to establish a correspondence between the result of DPL and of FDI, i.e. to find an expression for the effective force exerted on the sample surface during DPL. The conversion obtained by putting $D_{FDI}=D_{DPL}$ depends of course on the particular experimental set-up, but the order of magnitude and the procedure used to determine the correspondence have a general significance [263]. This effective force has only a parametrical mean and cannot be used to calculate the stiffness and to derive an approximated Young's modulus, because the Young's modulus of polymers depends on the frequency ω of the applied force. Such a dependence is usually represented

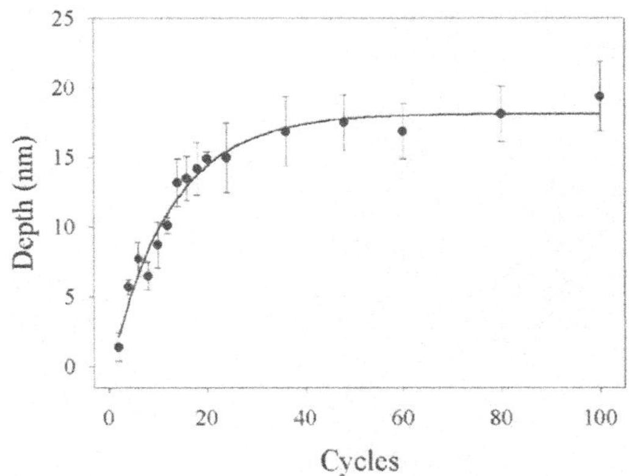

Fig. 27 Dependence of the depth of the carved holes on the number of cycles. The depth increases and reaches a plateau, whose height depend on F_{max}

through a complex modulus $E^*(\omega)=E_1(\omega)+iE_2(\omega)$ [269]. The real part E_1 (storage modulus) defines the energy stored in the specimen, the imaginary part E_2 (loss modulus) defines the dissipation of energy. At low frequencies, the polymer is rubber-like and has a low modulus E_1. At high frequencies, the polymer is glassy and has a larger E_1. At intermediate frequencies the polymer behaves as a viscoelastic solid, and its modulus E_1 increases with increasing frequency. The viscoelastic region is characterized by a maximum of the relaxation-time spectrum $H(\tau)$. $H(\tau)$ depends on the temperature and on the molecular weight and for highly dispersed polymers only a rough estimation of $H(\tau)$ can be provided. Both polymers employed in the present work have a glass temperature T_G of about 110 °C. So, at room temperature, they are in a glassy state both in FDI and in DPL, and their Young's modulus does not differ considerably in the two methods. However, for other polymers, or at other temperatures, these could be up to two orders of magnitude larger in DPL than in FDI. In particular, for PMMA, the viscoelastic region is localized between 10 and 10^4 Hz [270, 271]. The modification in FDI occurs at 1 Hz, whereas the modification in DPL occurs at $1.5{\cdot}10^5$ Hz, so that E is typically two orders of magnitude larger in DPL than in FDI.

Hence, when a hole of the same depth is obtained with the two methods, the force applied in DPL could be much greater than the force applied in FDI. The Young's modulus at higher frequency cannot be measured with force-displacement curves in an approximated way because no force-displacement curves can be acquired at this frequency.

Figure 28 shows the dependence of V_{pos} and V_{neg} on F_{max} for holes obtained by means of FDI. Also in this case the volumes follow a cubic law.

For the nanoindented holes, V_{pos} and V_{neg} are quite the same, whereas, for the holes carved by means of DPL, V_{pos} is bigger than V_{neg} (see Sect. 6.2.3). When the volumes are plotted vs the effective force calculated by means of the depth of the holes, the two plots of V_{neg} overlap, but not the two plots of V_{pos}.

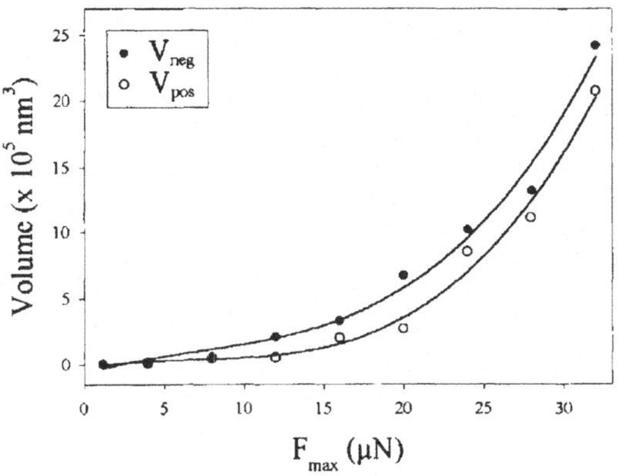

Fig. 28 Dependence of the volume of the holes V_{neg} (*full circles*) and of the border walls V_{pos} (*open circles*) on F_{max}. Both volumes follow a cubic law. For F_{max}<1 μN no modification is obtained. For all larger maximum forces the two volumes are approximately the same

It has been proved that the increase of volume ($V_{pos}>V_{neg}$) is due to the particular lithography method [263]. Holes have been carved by indenting the sample up to a fixed force F_{max} (as in FDI), but adding an oscillation of increasing amplitude. The volume of the border walls increases with the amplitude of the added oscillation, whereas V_{neg} stays nearly constant. Only for small amplitudes of the added oscillation the two volumes are the same. This experiment proves that the fast oscillations of the tip in DPL are responsible for the production of a border wall, whose volume is bigger than the volume of the holes. When such oscillations are not present, as in FDI, $V_{neg}\cong V_{pos}$, i.e. the material that makes up the border wall is simply the material that has been carved out from the holes by the tip. When, as in DPL, fast tip oscillations are present, at least a change of the density of the polymer must be assumed, in order to explain the increase of the volume. Such a change of the density of the polymer could in turn be caused and/or accompanied by changes in the free volume of the polymer or by chemical reactions of the polymer chains.

6.4
Insights into the Process of Surface Modification

In order to characterize the material in the border walls and to study its physico-chemical properties, force-displacement curves have been acquired on a surface where rectangles with large border walls, like the second square in Fig. 25, had previously been written [272].

Figure 29 shows the result of such a measurement. Figure 29a is the topography of the lithographed surface, acquired with force-displacement curves. In the bottom part of the figure, three typical force-displacement curves can be seen, acquired at the locations indicated in Fig. 29a. Approach (withdrawal) curves are plotted with full (open) circles.

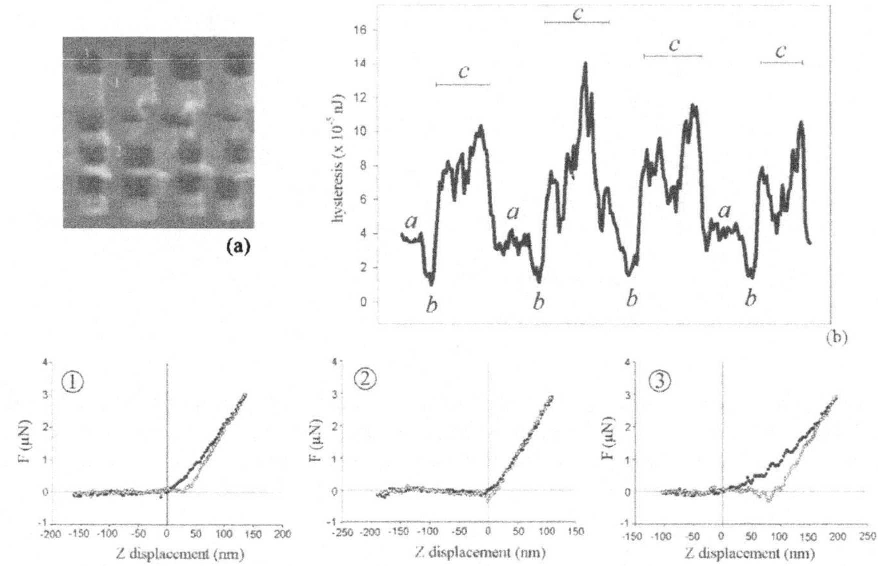

Fig. 29 a Topography acquired through force-displacement curves of a PMMA sample, where 16 rectangles with large border walls had been previously written (scan width 7 μm, *z* scale 350 nm, 200×200 pixels). The *numbers* show the location of the force-displacement curves plotted in the bottom of the figure. The *white line* shows the location of the points on which the hysteresis has been calculated. **b** Hysteresis between the contact lines of the approach and of the withdrawal curve. Regions (*a*) are the unmodified surfaces, with small hysteresis, in regions (*b*) the polymer is compressed and the hysteresis is very small, in regions (*c*) (residues of the border walls of the single holes inside the squares or large border walls outside the squares) the hysteresis is very much larger than in the other regions. In the *bottom part of the figure*, three curves representative of the three regions are plotted: a curve on unmodified polymer (1), a curve on compressed polymer (2), and a curve on border walls (3)

The first curve (1) was acquired on unmodified PMMA. The most important feature is the hysteresis between the contact lines, due to the plastic deformation of the sample [6, 273]. The importance of this quantity can be understood by comparing this curve with the second curve (2), acquired at the left end of a rectangle, i.e. where no residues of the border walls are present. In this curve, no hysteresis is present. This means that the sample behaves ideally elastic. An ideally elastic sample, even if it undergoes a deformation during the approach, regains its shape step by step during the withdrawal, so that the loading and unloading curves, i.e. the approach and withdrawal contact lines, overlap. On the other hand, if a sample is not able to regain its shape, at a given displacement the unload forces are smaller than the load forces. Unmodified PMMA is plastically deformed by the tip during the loading curve, whereas the PMMA at the left end of the rectangles, already compressed during DPL, cannot undergo a further plastic deformation. Force-displacement curves are able to sense the difference in density between the two regions. The compression, and hence the hysteresis, becomes larger on the residues of the border walls inside the rectangles (curve 3).

Also the curvature of the loading curve is an indication of the presence of a compressible material. The third curve in Fig. 29 can be fitted with a function in the form $F=F_0+az^2$ without any linear term, whereas the second curve can be fitted with a linear term. This means that, when the tip is placed on the residues of the border walls inside the rectangles, it pushes at the beginning on a soft material that becomes stiffer and stiffer as the tip compresses it. Consequently, the slope of the loading curve goes from 0 (very soft material) to k_c, i.e. the elastic constant of the cantilever (very stiff material).

The unloading curve always has a slope equal to k_c. This means that the low density material (residues of the border walls inside the rectangles or large border walls outside the rectangles) has been completely compressed, the tip has indented it, carving a hole, and has reached the underlying unmodified PMMA. Therefore two parts of the contact region can be distinguished: a part with hysteresis, corresponding to the indentation of the low density material and showing a plastic response, and a part without hysteresis, corresponding to the contact with the unmodified PMMA and showing an elastic response.

Figure 29b shows the work of plastic deformation A_p, or hysteresis of the contact lines, i.e. the area between the approach and the withdrawal contact lines, along the white line of Fig. 29a. Three different regions can be distinguished: the unmodified PMMA (*a*) with $A_p \cong 4\times10^{-5}$ nJ, the compressed region at the left end of the rectangles (*b*) with $A_p \cong 1\times10^{-5}$ nJ and the low-density PMMA (*c*), i.e. residues of the border walls inside the rectangles or border walls outside the rectangles, with 6×10^{-5} nJ$<A_p<14\times10^{-5}$ nJ.

Several approach curves, e.g. curve 3, present a periodic curve superimposed on the contact line. Such a modulation has been interpreted through a simple model [272]. The border wall can be seen as a pile of globular PMMA cluster-like structures. At the beginning of the loading phase, the tip pushes on one of these globular structures, and the force increases while the tip deforms and/or removes the cluster. Once the cluster has been removed, the force decreases until the tip meets another particle. This process is repeated several times. Also withdrawal curves on high border walls show a periodic curve superimposed on the unloading curve [272].

Several other proofs of the different density of the modified and unmodified polymer have been collected [263, 272]:

1. Modified and unmodified PMMA have different adhesion with the tip in air. The larger values of the adhesion on the low-density material are due to the different stiffness of the polymer structures. Since border walls and border wall residues are very soft, they are able to wrap up the tip, increasing the contact area and the adhesion.

2. When force-displacement curves are acquired with $F_{max}<1$ µN, holes are carved only on the border walls. This is due to the different hardness and to the different density of the polymer.
3. In contact mode images in water only the border walls are dragged away, because they are softer and not very well anchored to the bulk PMMA.

Border walls can be selectively solved by dipping the sample in a solution of acetone in water for about 10 min. This is due to the fact that the border walls are not very well anchored to the bulk PMMA and to the lower density of the border walls, allowing the solvent to penetrate into the polymer. This procedure has been used for the fabrication of sub-micrometre structures of the ferroelectric copolymer polyvinylidenefluoride/trifluorethylene (P(VDF-TrFE) (for more details on the material and the sample preparation, see Sect. 7.6). The polymer was lithographed with a "chess board"-like structure, where the white fields consist of "pulse-trains" producing overlapping holes. Figure 30a shows the result. For "developing" the polymer film, in the sense of common photolithography, the surface is subsequently to DPL immersed in methylethylketone for 10 s at room temperature. Imaging the same area afterwards reveals that only the modified areas are selectively dissolved (Fig. 30b). Figure 30c shows the cross sections (black lines in Fig. 30a,b) of the topographic images before and after the development step. The depth of the etched areas equals the film thickness (~70 nm) measured by scratching the film before. Therefore, we conclude that DPL is able to etch the film down to the substrate. For testing the size limits for the smallest structures, we reduced the scan area and used a mask containing isolated features. In Fig. 31 the smallest fabricated polymer dots, having a full half width diameter of around 250 nm, are shown. Further size reduction failed due to the bad adhesion of the polymer dots to the underlying gold substrate. Such a micromachining of three-dimensional polymeric structures can be important for the application in micro-electro-mechanical systems (MEMS) or sensors and has attracted much interest in the last few years [274–276].

DPL causes changes not only in the density of the lithographed polymer, but also in its chemical nature. In order to confirm this supposition, measurements using both SFM and other techniques have been performed.

Differences in the chemical nature of regions of the sample can be highlighted through measurements of the adhesion in water (see the review article by J. Vancso, contained within this volume). The adhesion in water, measured on lithographed squares like the ones in Fig. 29 is not influenced by the meniscus force [6] and, since the interaction force is very much smaller and the tip does not need to indent the surface, it is also not influenced by the density of the sample.

The adhesion on modified and unmodified PMMA in water is the same. Since border walls are quite flat, cross-talk phenomena between adhesion and topography [6, 277, 278] are observed only on the border of the squares and of the border walls, where the steep edges lead to a change of the contact area and hence of the adhesion.

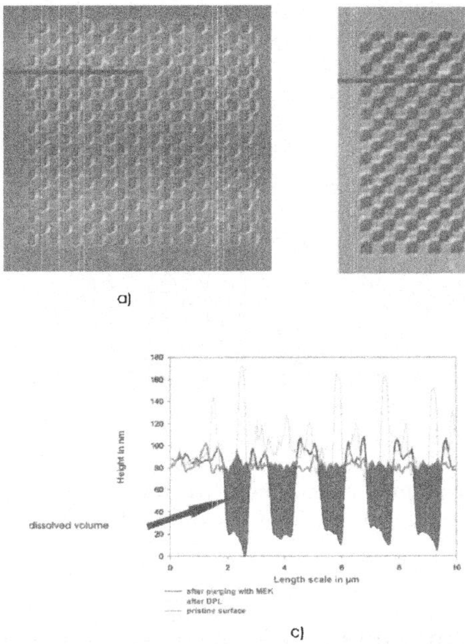

Fig. 30a–c Topography: **a** before; **b** after developing the polymer film by immersing the surface for 10 s in methylethylketone at room temperature; **c** cross sections of both topographies (*black lines*) reveal film etching down to the gold substrate (scan width 20 µm, *z* scale 166 nm (**a**) and 175 nm (**b**), 500×500 pixel)

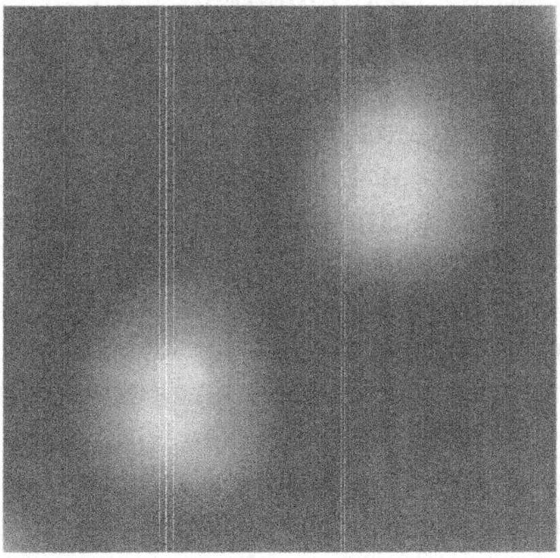

Fig. 31 Topography of the smallest fabricated polymer dots (scan width 970 nm, *z* scale 35 nm, 100×100 pixel). The full half width diameters are ~250 nm

Figure 32 shows the histograms of the jump-off-contact force on modified and unmodified polystyrene (PS). It is evident that the jump-off-contact force on modified PS is smaller than on unmodified PS (the mean force is 5.8 nN for unmodified PS and 2.3 nN for modified PS).

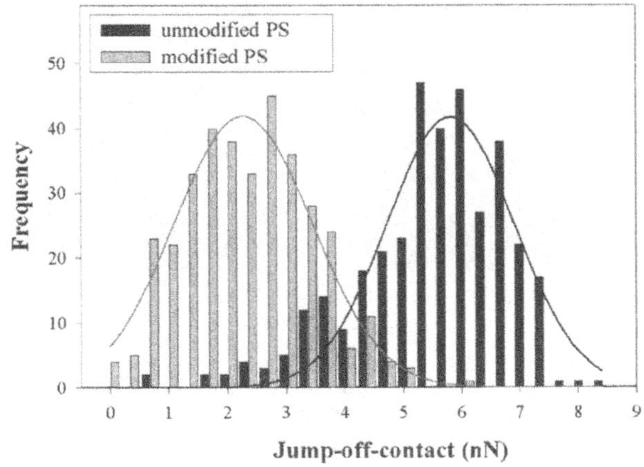

Fig. 32 Jump-off-contact force on previously lithographed PS. The adhesion on modified PS is smaller than the adhesion on unmodified PS, due to the presence of hydrophilic groups created by DPL

The smaller adhesion on modified PS can be explained only through a change in the hydrophilicity of the polymer, caused by chain scission followed by oxidation. During DPL, the fast oscillating tip repeatedly contacting the surface is able to break the chains. Such a chain scission leads to the creation of hydrophilic groups that change the total hydrophilicity of the sample. On PMMA, that is already hydrophilic, the number of new hydrophilic groups created by DPL represents a small percentage of the total amount of hydrophilic groups and the adhesion of PMMA to the tip in water, that is already very small, becomes certainly smaller, but the slight difference between the two histograms does not permit them to be sorted out. On the other hand, PS is hydrophobic, and the small number of hydrophilic groups created by DPL has a dramatic effect on the total hydrophilicity of the sample surface, further enhanced by the fact that the new hydrophilic end-groups will dispose themselves on the surface, in contact with water. This means that on unmodified PS the tip comes in contact with a hydrophobic surface, leading to a particularly large adhesion [6, 279–282], but on modified PS and on border walls it comes in contact mainly with hydrophilic end-groups, and this leads to a smaller adhesion. The difference is large enough to single out the two histograms.

The comparison between the two techniques has shown that DPL is more flexible than FDI (any kind of pattern can be transferred on the surface) and less time consuming. On the other hand, very high and large border walls are produced in DPL. The analysis of the properties of these border walls reveals that DPL causes changes in the physical (density) and chemical (length of the chains) properties of the polymer.

FDI is preferable to DPL when a modification of the topography of the sample with a simple pattern is desired. However, DPL has a lot of advantages when the pattern is more complicate. Additionally, DPL can be used in order to change locally the physico-chemical properties of the sample, and to obtain, for instance, structures with smaller density than the unmodified polymer, or hydrophilic structures in a hydrophobic sample.

7 Materials Contrasts of Electric Properties

7.1 Motivation and Introduction

Experiments on electrical materials contrasts are important because very often the mechanical properties of two polymers are similar and the polymers cannot be distinguished by the use of mechanical contrasts, e.g. stiffness and friction. Since classical polymers are generally insulators, one would expect SFM-measurements of the electrical properties of polymers to be focused just on contrasts of the dielectric constant (DK). However, the assumption that only the dielectric properties of polymers are analysed with SFM techniques is not true. Of course, there are still applications of new polymer insulators, e.g. the low-DK materials for microelectronic devices, requiring measurements of the dielectric constant on a small scale, but conductivity measurements by means of SFM are essential for new polymeric conductors as key components of organic solar energy conversion cells and organic light emitting diodes (oLEDs). Hence, lots of SFM methods to determine electrical properties, such as conductivity, dielectric properties, charge build-up or charge migration, polarisation, carrier injection etc., are already important in the field of polymer analysis. Nanoscale materials are mainly defined by the interface and its properties, which makes a microscopy technique essential for their investigation. Hence, electrical methods in SFM are of increasing importance. The very complex charge transfer in nanoscale materials is based on incoherent carrier transport mechanisms across and along the interfaces, and it determines the macroscopic properties of the nanoscale materials [283].

Using an SFM-type microscope, measurements of electrical properties have to be performed either in contact mode, i.e. with the conductive SFM tip being in mechanical contact with the surface, or in non-contact mode, a dynamic mode with vibrating cantilever, or in the so-called lift mode, where a line is repeatedly scanned with a chosen and controllable tip-surface distance. Contrasts of capacitance, dielectric constants or potentials can be achieved both in the contact and in the non-contact mode, and the quality of the result is not influenced by the sensitivity of the chosen instrument alone. All the sample properties (most of all sample thickness and size) affect the sensitivity of both methods, so that the prediction of the most successful technique is not always possible.

The overview of the literature [284] should start with polymers and polymer-related materials like composites. However, for electrical properties a lot of pioneering experiments were done in the field of semiconductors, ceramics or carbon nanotubes. So, even if we would like to focus the literature review on polymers, the semiconductor-based problems and the industry behind semiconductors are the driving motor for nearly all experimental and theoretical development over the last decades in this SFM field and cannot be ignored [285–288].

After an overview over the experimental techniques and results from the literature (Sect. 7.2) and some words about technical aspects and our experience concerning problems with some materials (Sect. 7.3), the experiments of the authors can be outlined as follows: first, measurements of ohmic and capacitive currents in the contact mode are described (Sect. 7.4), followed by a description of some surface charge measurements in the non-contact mode (Sect. 7.5). The chapter closes with some experiments to probe electro-mechanical properties by the use of piezo response microscopy (Sect. 7.6) with its own brief literature overview. All three experimental parts are opened by a short introduction to the SFM techniques implemented in our lab.

7.2 Literature Overview

The first and best known near-field technique to measure electrical properties in the nanoscale is of course Scanning Tunnelling Microscopy (STM). Since its invention by Binnig et al., STM has been used to explore the mechanisms of lots of phenomena on surfaces [289–294], ranging from experiments concerning the local work function to the use of an STM-tip to induce electropolymerisation [295]. Most of all, STM provides us with atomically resolved images of the surface structure.

The problem with STM appears if the surface is contaminated with insulators or if the sample is only partially conductive. Conductivity SFM (c-SFM) has the following main advantage over an STM: the feedback does not depend on a tunnelling process through a conductive surface but only on repulsive forces between a tip and any solid surface. For c-SFM measurements, an electrical circuit is closed between a conductive tip and cantilever, a voltage source (a.c. or d.c.), a current measuring unit (preferably a current-to-voltage converter), the back electrode of the sample and the sample itself, on whose surface the tip is performing a scan. The disadvantages of c-SFM with respect to STM are the lower achievable scan speed, the higher sensitivity to mechanical noise and the increased complexity of the experimental set-up. A special development in this field is the combination of an STM and an SFM feedback, implemented by Battiston et al. [296], where a fuzzy logic controller balances between an STM constant-current feedback and a non-contact SFM constant-distance feedback to get rid of problems at non-conductive sites.

c-SFM has been used for the analysis of semiconductors or related topics such as quantum dots [297–309], metals, electrodes and metal filled membrane pores [310–318], biological systems [319–322], organic and polymeric systems [323–325] as well as inorganic materials like carbon fibres, carbon nanotubes, carbon black and conductor-insulator composites [326–332]. The experiments were inspired or complemented by theoretical work and tip-sample analysis [333–338].

In 2000, De Wolf et al. [339] gave a status report and review of two-dimensional carrier and dopant profiling with semiconductors using scanning probe microscopy, additionally comparing scanning spreading resistance microscopy (SSRM) [340–343], a technique similar to c-SFM, to scanning capacitance microscopy (SCM). SCM started in the 1980s [344–350] by using a capacitive sensor originally constructed for videodisc players. Three approaches to measure the capacitance can be realised:

1. A capacitance sensor [351], in general based on the resonance-peak-shift principle, is connected to the conductive SFM tip.
2. The imaginary part current passing the sample is measured by a lock-in amplifier (see Sect. 7.4.6).
3. A dynamic mode experiment is performed by exciting the cantilever with an external voltage $U_\omega \sin(\omega t)$ and $F_{2\omega}$ is measured.

With SCM, sensitivity and resolution on the time scale are more than sufficient to probe the influence of carrier injection and space charge relaxation in semiconductors and conducting polymers. It turned out that, for semiconductors and therefore for polymers, the d.c. probing voltage in c-SFM or SSRM may easily change the space charge distribution or polarise the electrodes, leading to a modification of the carrier injection (self-blocking) and to the migration of the dopant or of the traps. This was proved by Born and Wiesendanger [352], who for an n-doped silicon with p-doped stripes showed the significant changes of space charge distribution after only one scan with a d.c.-biased back electrode. The destructive effects of the d.c. voltage can be avoided as shown by Tomiye and Yao [353]. Both results are important for polymer investigations due to the similarities in the band structure and the influence of space charges and traps on the electronic properties.

The best resolution in SCM today with zeptofarad (10^{-21} F) sensitivity was realised by Tran et al. [354]. Although the dielectric properties of surfaces can be studied with SCM, the research in this direction has been limited to thin layers of insulating materials on conducting substrates.

The term "electrical force microscopy" or "electrostatic force microscopy" (EFM) is used if the detection of the electrical properties like surface potential or charge is based on a force, leading to a d.c. cantilever bending or to a change in the vibration amplitude or frequency of a vibrating cantilever.

Since Coulomb forces decay much more slowly than van der Waals forces, an EFM experiment can be performed even without any feedback, if the sample roughness is low compared to the tip-sample distance. Terris et al. used this technique for the first time to investigate a contact electrification of a metal-insulator system [355–357]. Later, further improved experiments on contact electrification with single charge sensitivity were performed by

Schönenberger and Alvarado [358–360] and Sugawara et al. [361]; the dynamic of the charge spreading was analysed by Wintle [362].

The Kelvin method was first described in the year 1898 by Lord Kelvin [363] and in 1932 Zisman [364] added the a.c. technique. SFM techniques based on the Kelvin method were reviewed 1997 in the article by Steinke et al. [365]. Kelvin probe detection approaches in an SFM [286, 366–377] are based on non-contact experiments using vibrating cantilevers to measure the work function or the contact potential between a conducting specimen and the vibrating SFM tip. In such an experimental set-up, the conductive tip and the sample surface represent a capacitor with varying distance. Both electrodes are connected in series via a sensitive a.c. current meter (preferable a fast current-to-voltage converter) and a variable d.c. voltage source to supply a compensation voltage. In case of an insulator sample placed on an electrode, a dielectric material is added to the circuit and the surface and space charges play the role of an additional contact potential. The total d.c. voltage between the electrodes is the sum of these voltages. If the SFM tip electrode moves periodically up and down with a frequency ω, the corresponding modulation of the capacity C leads to a displacement current I which is proportional to $(\partial C/\partial t)$. The displacement current disappears if the d.c. potential between the electrodes is zero, i.e. if the applied voltage compensates the surface potential. In this case there is no electric field inside the capacitor, and the value of the compensating voltage is equal to the negative value of the surface potential.

Moreover, in an SFM in the non-contact mode with a vibrating cantilever driven by a dither piezo at the resonance frequency, the topography feedback which is based on the damping of the cantilever vibration or the distance-dependent frequency shift can be used to keep the average distance between tip and surface constant, providing topography images simultaneously. Additionally, the compensation with the d.c. voltage can be performed automatically with a second feedback unit, so that during the scanning of the SFM tip over the surface it is possible to generate images of the contact potentials or charges. A non-calibrated materials contrast, free of the electronic noise of such a second feedback amplification, can be achieved by just measuring the a.c. currents [378].

The detection and compensation of the a.c. current is the classical Kelvin method; however, the resulting electrostatic forces, i.e. the corresponding cantilever bending, can also be used to establish a potential sensitive feedback. If an a.c. voltage is applied between the tip and the back electrode of the sample instead of using the dither piezo, the Maxwell stress microscopy (MSM) [379–381] or the electrostatic force microscopy (EFM) [317, 382–393] can be performed.

The total electrical potential U is given by

$$U = U_{DC} + U_{\omega}\sin(\omega t) = \phi + U_{ext} + U_{\omega}\sin(\omega t) \tag{17}$$

where Φ is the surface (contact) potential, U_{ext} an external d.c. bias enabling a Kelvin-type compensation and $U_{\omega}\sin(\omega t)$ an external excitation voltage. The attractive force F on the cantilever can be written in the form

$$F = -{}^1/_2\, U^2(\partial C/\partial z) = -{}^1/_2\,(U_{DC}+U_\omega \sin(\omega t))^2(\partial C/\partial z) \qquad (18)$$

where C is the capacity, and can be separated in three frequency components F_{DC}, F_ω, and $F_{2\omega}$ as follows:

$$F_{DC} = -{}^1/_2 U_{DC}^2(\partial C/\partial z), F_\omega = -U_{DC}U_\omega \sin(\omega t)(\partial C/\partial z) \text{ and}$$
$$F_{2\omega} = -{}^1/_4 U_\omega^2(1-\cos(2\omega t))(\partial C/\partial z) \qquad (19)$$

Thus, the surface potential Φ can be directly deduced from the cantilever bending measured at the frequency ω with the help of a lock-in amplifier, or by implementing a Kelvin-type compensation feedback using $U_{ext}=-\Phi$, i.e. $U_{DC}=0$ and $F_\omega=0$. Several additional experimental set-ups can be realised. For example, if the sample surface is not an insulator but electrically conductive or without any materials contrast giving rise to surface charge build-up, $F_{2\omega}$ depends only on the tip-sample distance and the dielectric constant and charge of the capacitor, so that a topography feedback for MSM can be realised by using the cantilever deflection measured at 2ω. In this case, 2ω should be adjusted to the resonance frequency ω_{res} of the used cantilever. Furthermore, if a standard non-contact set-up with a dither piezo for a cantilever excitation at ω_{res} is used and $\omega \neq \omega_{res}$ and $2\omega \neq \omega_{res}$, frequencies for the electrical excitation can be found where a cantilever vibration can be demodulated from the deflection signal. Hence, contrasts of the surface potential and the capacitance can be deduced [394, 395]. However, the sensitivity depends on the Q-factor of the lever at the chosen frequency and Q is small in the off-resonance states outside the different vibrational modes of the cantilever.

If the time consumption is acceptable and the image drift is negligible, a scan line can be scanned twice to separate topography and electrical properties. In this case, a first scan in contact or better in a dynamic mode without an electrical excitation is performed. The tip is lifted and for the following second scan the z-piezo is controlled in a way that the tip follows the same topography as for the first scan (constant tip-sample distance or "interleave" scan). During this second line scan, one of the above-mentioned measurements of electrical properties can be performed [396].

Using a cantilever with ω_{res} as a mixer, a two-frequency stimulation with ω_1 (e.g. an a.c. potential at a surface electrode) and ω_2 (e.g. a stimulation with chopped light) can be performed by choosing $\omega_1-\omega_2=\omega_{res}$. With this or some other heterodyne technique, experiments on the time scale of today's processors can be performed [287, 397–403].

Some keywords for further reading on measurements of electrical properties using the SFM are: pulsed-force mode [33, 404, 405], special SFM tips [406–409], lithography assisted by electric potentials [244, 254, 410–426], measurements of polarisation and water droplet manipulation [427–429] or reading and writing of ferroelectric domains and piezoactivity [393, 430–438].

7.3 Technical Aspects and Technological Problems

The drawback in the experimental set-up of all electrical measurements with the SFM is mainly due to the unshielded SFM tip and cantilever, regardless whether contact or non-contact approaches are used. Interesting solutions [439] for shielding the electrode are not yet on the market.

In order to be able to state that a conductive probe works, it should be stable over at least several hours. This is principally a question of the wear resistance of the conductive coating and additionally of its ability to withstand high current densities or the resulting temperature. Within the production process one has to ensure that the very last nanometers of the tip are conductive [440]. Conductive diamond and other commercially available hard coatings (tungsten carbide, W_2C, and titanium nitride, TiN), as well as some other metals evaporated in our in-house production, are used in our studies.

Figure 33 gives an overview of the upcoming problems with evaporated metal coatings, in this case a gold coating with a thin chromium sub-layer on silicon nitride. The temperature history during the evaporation process can free inherent stress in the cantilever, as seen in Fig. 33a. The result may be a formidable bending. Additionally, the mismatch of the thermal expansion coefficients can bend the lever if the coatings do not have the same thickness on both sides, which in turn can be easily assured by rotating the cantilever during the evaporation. The metallisation may be under stress if no suitable temperature program and choice of the thickness was used; this stress may cause partial delamination of the coating (Fig. 33c). If, during a conductivity measurement, the current density is too high, the gold coating may even melt (Fig. 33b). This happens rather often because increasing the excitation voltage is a well known way to increase the signal-to-noise ratio. In the long run, mechanical loading will abrade the conductive coating (Fig. 33d), and during this it is not easy to distinguish between the time dependency of the conductivity of the sample and that of the conductivity of the tip.

A deep insight into the problem of contact mechanics involved in a conductivity measurement using an SFM tip can be found in the paper by Lantz et al. [441]. In this article the contact area was derived for the case of ohmic contacts using the Maugis-Dugdale model [104] (see Sect. 2.1). However, the uncertainty is still related to the problem of the conductivity of the tip apex. If a sharp tip is not absolutely necessary, a possible solution to this problem is to add electro-chemically a copper layer to the chromium sub-layer (Fig. 33e,f).

Additionally, problems related to resolution and reliability can occur from the variation of the contact area. Great care has to be taken to optimise the feedback parameters, so that the indentation depth of the tip is as constant as possible. A recent publication [442], focussing on scanning capacitance microscopy, adds to this list the problem of adsorbed water, present at least at elevated relative humidity. Due to the high dielectric constant of water

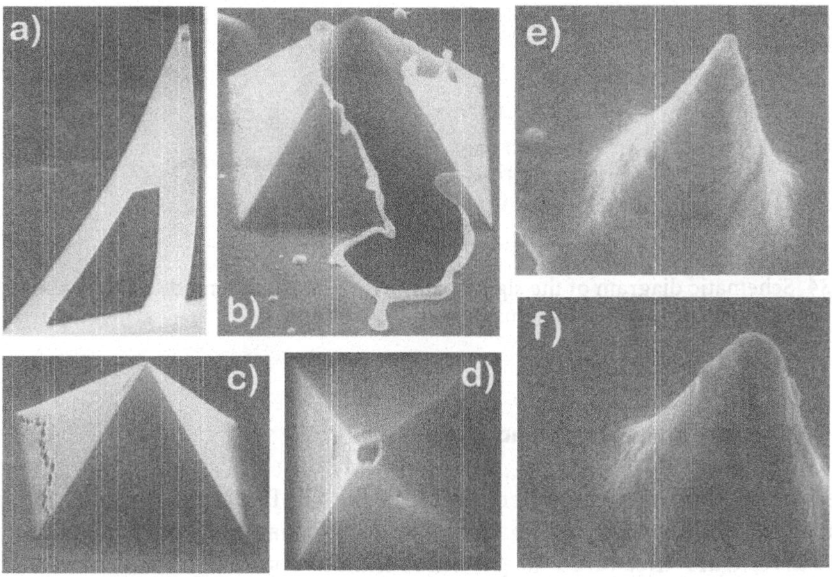

Fig. 33a–f SEM images of metal coated cantilevers and tips (silicon nitride **a–d**, silicon **e,f**) showing different failures. The base of all tips is ~4×4 μm: **a** cantilever bending after evaporation with Cr and Au; **b** molten gold (melting point: 1065 °C) after an excessive current density during a conductivity experiment on a carbon fibre (melting point: 1065 °C); **c** delamination of the metal coating due to mismatching thermal expansion; **d** wear of the metal coating; **e,f** electrochemically copper coated silicon tips with chromium sub-layer with different copper thickness

(ε=80), the resolution and sensitivity are significantly altered. Taking into account that even a small contamination of the surface can lead to a high ion concentration, it is evident that a scanning tip has to be regarded as an electrode surrounded by an electrolyte with a high dielectric constant.

The following sections will give an overview of our experiments with some partially conducting and some semi-conducting samples, measured either in the contact mode (Sect. 7.5) to obtain ohmic and capacitive currents or in the non-contact mode (Sect. 7.6) to get information about the distribution of surface charges.

Kelvin microscopy will not be presented within these studies. The important drawback with Kelvin microscopy is, that a second feedback is used to compensate the d.c. potential between surface and tip. In the case of surface or sub-surface charges on a polymer insulator, the corresponding signal is small and a high gain is needed, though the errors within the topography feedback induce much more pronounced responses as the electric potential itself. In other cases (see Sect. 7.5.2.), the surface potential was much too high to get even a stable van der Waals-dominated non-contact feedback.

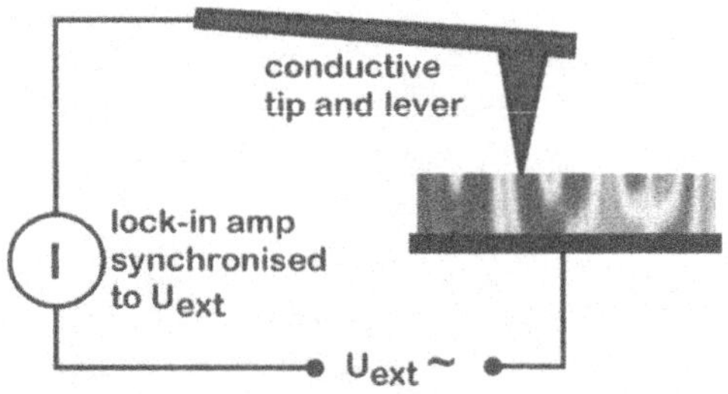

Fig. 34 Schematic diagram of the signal pathway in the contact mode SFM, enabling current imaging of the real and imaginary part

7.4 A.C. Conductivity in the SFM Contact Mode

The first method presented here employs the SFM in contact mode. Simultaneous with the topography image the real and imaginary part of an alternating current are measured to get information about the local conductivity and the dielectric constant. The SFM is provided with a current measuring circuit (Fig. 34), consisting of an alternating voltage source and a current pre-amplifier, and is closed over a conductive tip in contact with the sample surface.

The amplified signal is fed into a lock-in amplifier, whose phase is synchronised to the voltage source in such a way that real and imaginary parts represent the ohmic and the capacitive current flowing through the sample. The voltage outputs of the lock-in amplifier are analysed after analog-to-digital conversion as an image. It is worth explaining the phase compensation procedure and the problem of parasitic capacitive currents in more detail. The experiment starts with the tip in contact without an external potential, in order to adjust the height position of the z-piezo. Afterwards the tip is lifted in such a way that it just snaps off from the surface. After that, the tip can be moved closer over a short distance in order to reduce the tip sample distance, but it should not touch the surface. After increasing the external a.c. potential to the desired value, the resulting current must be purely capacitive. Now the phase shift of the lock-in amplifier can be adjusted in such a way that the imaginary part is at its positive maximum or, in other words, that it reflects the current flowing through an ideal capacitor. If the tip is brought into contact again, the capacitive current changes and the in-phase current reflects the ohmic behaviour of the sample. Unfortunately, the tip and the cantilever are not shielded, which results in a high parasitic capacitance and thus a high background imaginary part. By feeding a compensating sinusoidal voltage into the subtracting input of the lock-in amplifier, or using an adjustable a.c. bridge circuit, the parasitic a.c. current can be subtracted and the lock-in gain can be increased to achieve a higher sensitivity.

Sample excitation with an alternating voltage instead of a d.c. voltage has several advantages:

1. With lock-in techniques, small currents are measured with a very high S/N-ratio, so that the current density in the tip zone can be minimised.
2. Blocking of the tip electrode by the formation of a barrier is negligible, because charge carrier accumulation is avoided. Such a blocking could decrease the injected d.c. current with time, giving an undesired instability and time-dependence.
3. The most important advantage is that two physically independent properties, related to the conductivity and the dielectric constant, are measured simultaneously, providing much deeper insight into the materials properties.

The signal-to-noise ratio of the system enables, using an a.c. excitation voltage, to reach the atomic resolution on highly oriented pyrolitic graphite (HOPG) in simultaneously measured images of force interaction and current [443]. Hence, this new technique opens a wide range of applications. Investigations of carbon fibres after single-fibre pull-out from polyphenylenesulfide [444], partially reduced barium tetratitanate ceramics [445], and conducting films plasma polymerised from 2-iodothiophene [446] are given as examples.

7.4.1
Graphite

Freshly cleaved highly oriented pyrolitic graphite (HOPG) was used to acquire atomic resolution pictures both for force and current interaction (Fig. 35). The mean value of 234 pm for the atom spacing is larger than the C-C binding in graphite (142 pm) and reflects the well known carbon site asymmetry: only every second carbon atom appears as a protrusion [447]. This result is only used to prove that the method reaches good stability and sensitivity. Since a detailed knowledge of the tip-surface distance or the contact area, not available in this experiment, is necessary for all known theoretical evaluations, it is not possible to state whether the measured current comes from a tunnelling process or from conductivity. The lifetime of the metal coated tip may be shorter than a few hours due to the high mechanical load and the high current density which are applied.

7.4.2
Electrically Conductive Ion Tracks

Diamond-like carbon (DLC) films are insulators. The transfer of the kinetic energy from a heavy ion breaking through this film leads to a small channel of molten and recrystallised carbon in the more stable form of graphite, which is electrically conducting. A possible application of these conducting ion-tracks within an insulator [448] is the generation of field emission [449–451], e.g. for displays or other vacuum electronic devices like hot filaments, because conducting tracks with a high aspect ratio lead to an enhancement of the local electric field at the film surface. Such DLC films can be produced

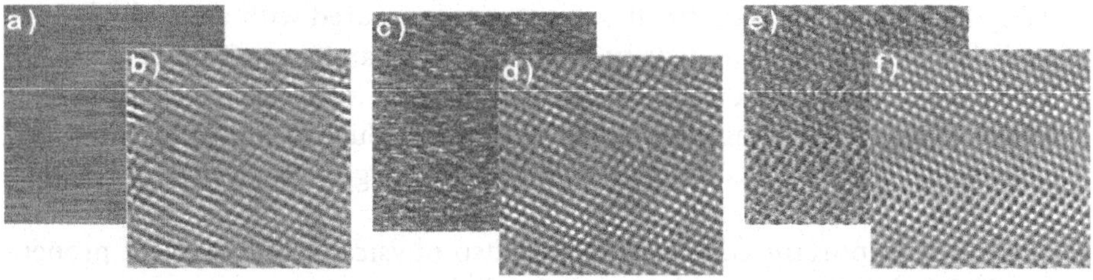

Fig. 35 Simultaneously measured: **a,b** topography; **c,d** repulsive force; **e,f** a.c. current amplitude on graphite before (**a**, **c**, **e**) and after (**b**, **d**, **f**) Fourier space filtering. The Fourier transform parameters for the inverse transformation of **b** are taken from the Fourier transformation of **e**. Scan width 5.5 nm, 500×500 pixel, scan speed 50 nm/s, repulsive force ~100 nN, a.c. excitation 3.9 mV at 102 kHz. In order to minimise the piezo amplifier noise, a weak feedback was used and the topography contrast is smaller than 100 pm. Force contrast ~1 nN, current contrast from 5.5 to 7 nA

either by ion beam techniques or by plasma deposition with magnetic filtering. For the film formation process, the carbon ions were implanted into the growing film with an energy of the order of 100 eV. The films are amorphous and contain 70–80% sp^3 bonds; the film thickness is 50 nm. The sample used was plasma deposited on highly doped silicon with the filtered arc method[1].

The heavy ion irradiation of the DLC films was performed[2] using Uranium projectiles of ~1 GeV with a dose of 10^{10} ions/cm^2, i.e. 100 ions per μm^2. The properties of these channels were studied by means of SFM using a conducting tip; both diamond and tungsten carbide coated tips[3] were used successfully.

Figure 36a shows the end caps of the graphitic filaments, looking like small hillocks due to a density and morphology change of the material at the ion impact site. The corresponding image of the a.c. amplitude of the tip collected current demonstrates the conductivity through the graphitic channel, which is contacted to the a.c. voltage source via the silicon substrate. The number of hillocks in the overview images corresponds to the applied ion fluence. The d.c.-current between the substrate and the SFM tip as a function of the voltage was measured by Krauser et al. [452] for a spot on an ion-track and a spot beside the track. Assuming a track cross section of 100 nm^2, the measured current at 5 V gave a current density of 10^4 A/cm^2 and an electrical resistivity of 50 Ωcm. Since the conductance of the tracks is four orders of magnitude smaller than that of crystalline graphite, the sample is a very imperfect graphite crystal or even partially amorphous carbon.

[1] Performed in the group of B. Schultrich, FhG Dresden, Germany. The sample has been put at our disposal by A. Weidinger, HMI, Berlin, Germany.

[2] At the UNILAC (Gesellschaft für Schwerionenforschung GmbH, Darmstadt, Germany).

[3] Supplied by Nanosensors, Germany, and NT-MDT, Russia, respectively.

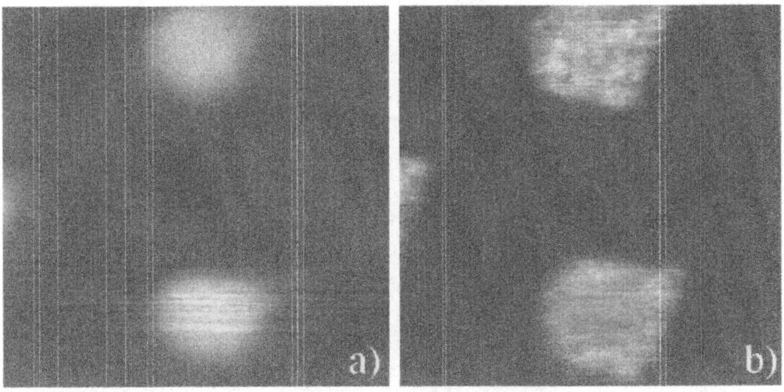

Fig. 36 a Topography. **b** A.c.-current amplitude of the irradiated DLC film. Each spot is a few nm in height and 10–20 nm in diameter. The excitation voltage is 100 mV at 51 kHz, the current is several 10 nA

In all experiments requiring conductive tips, such samples with small conductive channels can be used very easily to check the conductivity of the last few nanometers of an SFM tip.

Figure 37 gives an example of two different channels, both conductive, measured with a well conducting tip (foreground) and with a tip, which, due to mechanical wear or a high current density, had lost its conductivity at the very last 10 nm.

7.4.3
Carbon Fibre Surfaces

A carbon fibre (high-modulus type, diameter 7 μm) was cleaned with methanol and was embedded vertically for 105 μm in the zenith of a molten droplet (at 290 °C) of polyphenylenesulfide (PPS Fortron 205, Hoechst). After cooling to room temperature, the fibre was pulled out of the polymer droplet by a piezo translator (single fibre pull-out test), measuring the adhesion force with a force sensor [453]. The aim of the investigation on carbon fibre composites was to study interface failure mechanisms by debonding. The interface has failed at those sites, where the well conducting carbon fibre surface is free of polymer coating after the pull-out. If PPS sticks on the carbon fibre surface, the interface has a high adhesion strength and the failure is located within the polymer matrix. Since in Fig. 38d no polymer residues can be observed in the topography contrast after pull-out, their thickness must be less than some tenths of a nanometer. The fibrilled structure of the fibre does not differ remarkably from that of a clean reference fibre (Fig. 38a). The ohmic and capacitive current can be used to calculate amplitude and phase shift images, giving a better contrast of the remaining polymer. The image shows that even a thin isolating film leads to a phase shift of nearly 90° (Fig. 38c,f). The second row (Fig. 38b,e) shows an electrical heterogeneity on the partially conducting clean fibre surface. Grains in form of prolate ellipsoids on the parallel grooves can be found. After pull-out from PPS, the

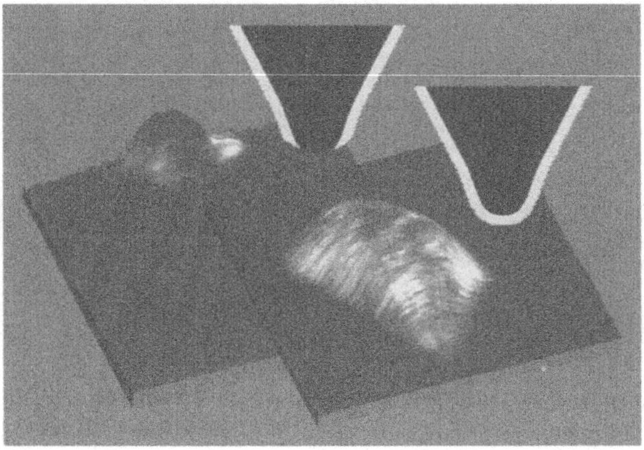

Fig. 37 Topography of two conductive graphitic channels, measured with an intact tip (*foreground*) and with a tip which has lost its conductivity at the very end. Scan width 65 nm, corrugation 3 nm. The local current density was encoded in *greyscale colour*, where *white* represents the highest current, and *dark grey* the current below the detection level. The excitation voltage is 100 mV at 51 kHz, the current is several 10 nA

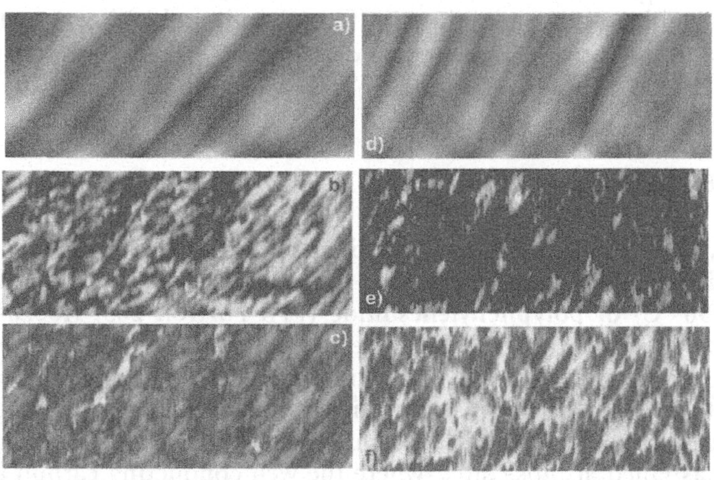

Fig. 38a–f Carbon fibre analysis: **a–c** before; **d–f** after a single fibre pull-out from polyphenylene sulfide. Topography (z scale: 72 nm (**a**) and 52 nm(**d**)), a.c. current amplitude ((**b**) and (**e**): 45 nA) and current phase shift ((**c**) and (**f**): 0° to 90°) are measured simultaneously. Excitation 2.58 mV at 102 kHz

darker zones and the corresponding brighter parts indicate a non-conducting surface coverage. On the other sites the interfaces had failed during debonding. The development of this technique led to an interesting experiment of interface thickness investigations on the same system, which is discussed in Sect. 4.5.2.

7.4.4
Polyaniline, a Conductive Polymer

The polyaniline sample[4] was doped in the microscope with a small droplet of hydrochloric acid (HCl 37%) for 20 min, followed by a drying period of ~1 h. The phase shift of the lock-in was adjusted as already explained to compensate the time lag of the electronics and the phase shift due to the parasitic capacitance. The measured conductivity and the capacitive current are volume properties. Since a conductive path to the back electrode is necessary to measure a real part current, non-conductive regions, which might be due to inhomogeneous doping or contamination, can isolate a conductive zone and prevent it from detection with this method. To learn about such complex heterogeneous samples as much as possible, different modes and material contrasts are useful, so topography, normal force (error signal), friction, and real and imaginary parts are measured for both the forward and backward scan. First of all, all materials contrasts in Fig. 39 show the region doped with the HCl. The conductivity (Fig. 39e) is higher than in the non-doped region (right part of the image), and the capacitive current is enhanced through the increase of the dielectric constant due to the water and/or HCl absorption. Also the friction is different on the doped and the non-doped regions. Additionally, some bright spots, i.e. very high ohmic currents, can be identified, as well as some weakly conducting sites. These sites can be attributed to non-conducting contamination particles (thickness influence on the capacity) and to inhomogeneity in the sample composition. Such inhomogeneity might be even a volume effect. The normal force shows how much the topography image can be disturbed by an imperfect feedback. The normal force signal can be used for analysing the quality of the capacitive and ohmic current images. It is evident that the inhomogeneity is not a feedback artefact but a real feature of the sample. Again, it is worth mentioning that the multiple contrasts, measured simultaneously, help to prevent the user from interpretation errors.

7.4.5
Plasma Polymerized Iodothiophene

Organic conductive films, e.g. polyaniline or polythiophene, have an interesting potential application due to their easy processability combined with a low weight. A plasma deposition process is even more interesting from a technical point of view, because it is, as opposed to wet chemistry, much more compatible with production processes in vacuum. The sample used was polymerized by Kruse et al. [454] on a silicon wafer over 20 min in a microwave plasma chamber at 2.45 GHz using 2-iodothiophene (at

[4] The polyaniline sample is kindly provided by Ralph Schmittgens, Departemento de engenharia metalurgica de materiais, Cid. Universitaria, Rio de Janeiro, Brasil. The synthesis was done at the Inst. de Fisica de São Carlos, Grupo de Polimeros Prof. Bernhard Gross, Brasil.

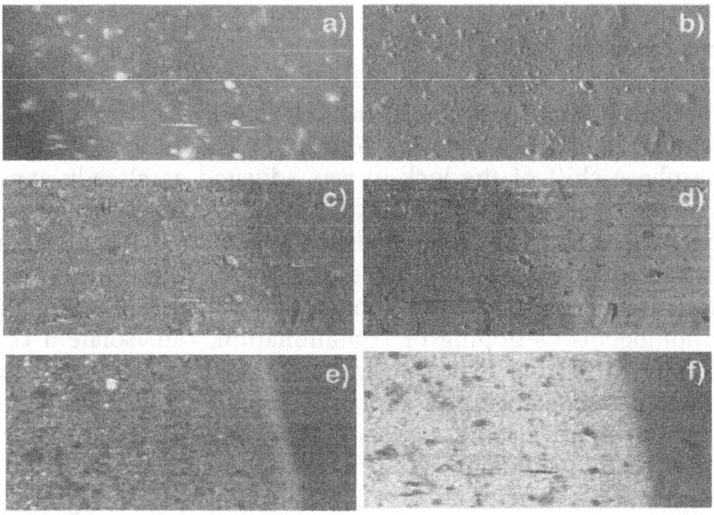

Fig. 39 a Topography. **b** Normal force error signal. Friction for: c the forward scan; **d** the backward scan; **e** real part (i.e. conductivity); **f** imaginary part (i.e. capacity) of the simultaneously measured a.c. current. The cantilever is made of silicon coated with diamond. Scan range 11.5×5.1 μm, topography corrugation 121 nm, force and friction in arbitrary units, ohmic current range 0–42 nA, capacitive current –60 pA to +30 pA, excitation 1 V at 16,789 Hz

0.065 mbar plus 5% Ar); the formed film had a conductivity up to 0.1 S/cm. The conduction mechanism is assisted by J^{3-}-ions (0.15%) and J^{5-}-ions (0.052%). Figure 40 (left) shows the topography of a plasma polymerised conducting film, 400–500 nm thick, on the right side the polymer film was removed from the silicon wafer substrate. A film consisting of round particles can be seen, whose size is between 50 and 900 nm and whose different mean thickness corresponds to the drop of the ohmic current (Fig. 40 middle). Figure 40 (right) presents the capacitive current. The particles and especially the film edge are surrounded by parallel patterns. Calculations of the two-dimensional local electric potential let us assume that this effect is to be attributed to the effect of the inhomogeneous contact area on capacity, and not to a contrast of dielectric constants. Additionally, the good conductivity of the doped silicon wafer and of the plasma polymer hindered us to get a high sensitivity for the 90° shifted phase signal of the imaginary part of the current, even by the use of background subtraction via the bridge circuit.

7.4.6
Barium Tetratitanate and Yttrium Doped Barium Titanate: A Partially Conductive and a Partially Semiconductive Ceramic

Electroceramics like barium titanate have an interesting industrial potential as ferroelectric permanent storage devices and as superconductors, respec-

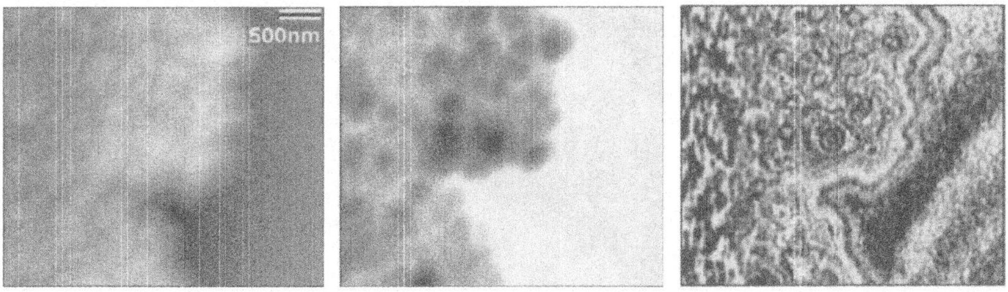

Fig. 40 Conducting film, plasma polymerised from 2-iodothiophene, on silicon. *Left*: topography contrast (shaded pseudo-3D-image) with 405 nm corrugation. *Middle*: real part (conductivity), with a contrast of 2.8 nA. *Right*: imaginary part (capacity), with a contrast of 270 pA, the a.c. currents shown in the *middle* and in the *right* image were simultaneously measured together with the topography. The cantilever is made of silicon nitride coated with gold. The excitation is 0.8 V at 60 kHz, the scan speed is 4.17 μm/s

tively. The analysis of electroceramics is an interesting test procedure for all kinds of electrical measurements using the SFM, because of their high dielectric constants and their high wear stability against the abrasion of the SFM tip.

One part of the investigated barium tetratitanate ($BaTi_4O_9$) ceramic[5] precursor (ε~35) was modified through an additional thermal treatment in a reducing atmosphere, which increases the dielectric loss from tgδ~0.0005 to 0.01 by forming conducting regions with Ti(III) predomination (from Ti(IV)). The topographic image of the ceramic sample (Fig. 41, right) does not show such heterogeneity and reflects only the polishing process. However, the middle image shows the ohmic current in a conducting region formed by the reduction of titanium valence state. The left image shows that these conducting regions have a higher dielectric constant, because therein the capacitive current is substantially higher. This correspondence has been found in each measurement. In some cases (see insets of current images, topography not shown) an additional interesting feature can be visualised: the edges of the conducting zones are brighter, i.e. better conducting than the inner areas. This effect can be attributed to a higher concentration of Ti(III) states and of carrier trap forming defects near the grain boundaries. Increasing the a.c. frequency leads to a dramatic decrease of the signal-to-noise ratio of the measured currents, because the current drops below the detection limit at about 5 kHz, demonstrating that the mobility of the carriers is strongly hindered. This leads to the assumption that ionic species are involved in the conductivity mechanism.

Since the yttrium-doped barium titanate sample[6] (Y-content 0.4%) is an n-type ceramic semiconductor, the conductivity increases under a negative

[5] The ceramic samples are kindly provided by V.P. Bovtoun and M.A. Leshchenko, Kiev, Ukraine.

[6] The ceramic samples are kindly provided by V.P. Bovtoun and M.A. Leshchenko, Kiev, Ukraine.

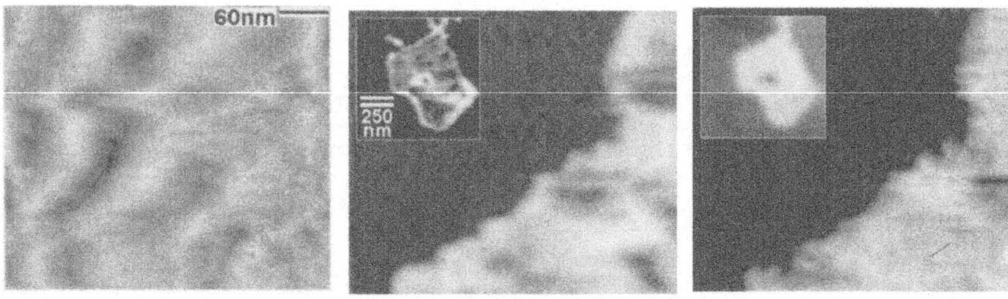

Fig. 41 Barium tetratitanate ceramics, partially reduced. The excitation is 5 V at 666.6 Hz, the scan speed 323 nm/s. *Left*: topography contrast with 4.8 nm corrugation. *Middle*: ohmic current, with a contrast of 170 pA (*inset*: 120 pA). *Right*: capacitive current with a contrast of 310 pA (*inset*: 810 pA)

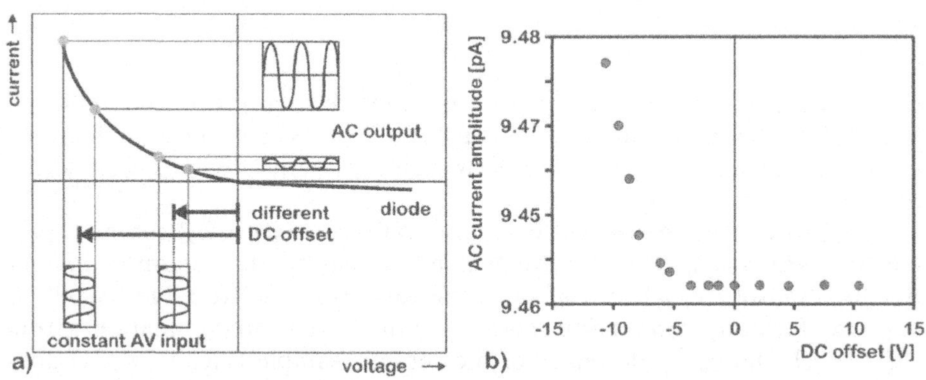

Fig. 42 **a** Schematic diagram of the result of an alternating voltage superimposed to a d.c. bias. **b** I/V-curve of a conducting region of yttrium doped barium titanate with constant tip position. The excitation is 0.5 V at 1 kHz

d.c. bias. Figure 42a shows the influence of a different d.c. bias voltage on the measured a.c. currents for a given constant excitation voltage. Probing a conductive site of the semiconductor ceramic at 1 kHz with a testing voltage of 500 mV, the d.c. offset was changed in the range of ±11 V, and this demonstrates the semiconductor behaviour of the sample (Fig. 42b).

The SFM images (Fig. 43) show, that using a d.c. offset of −10.57 V and an excitation voltage of 4.4 V at 1 kHz, high currents in the μA range can be measured. For this scenario, ceramics are ideal samples, because the low stiffness of the polymers makes it impossible to use excitation frequencies below the bandwidth limit of the constant-force feedback. The much stiffer ceramics permit one to perform measurements without cross-talk effects. Such cross-talk effects occur with soft polymer samples, due to a surface indentation by the tip, which is at virtual ground and so attracted by the back electrode under potential.

The yttrium-doped barium titanate shows a composition similar to a slice-on-slice morphology. It is not definitely clear whether some of the

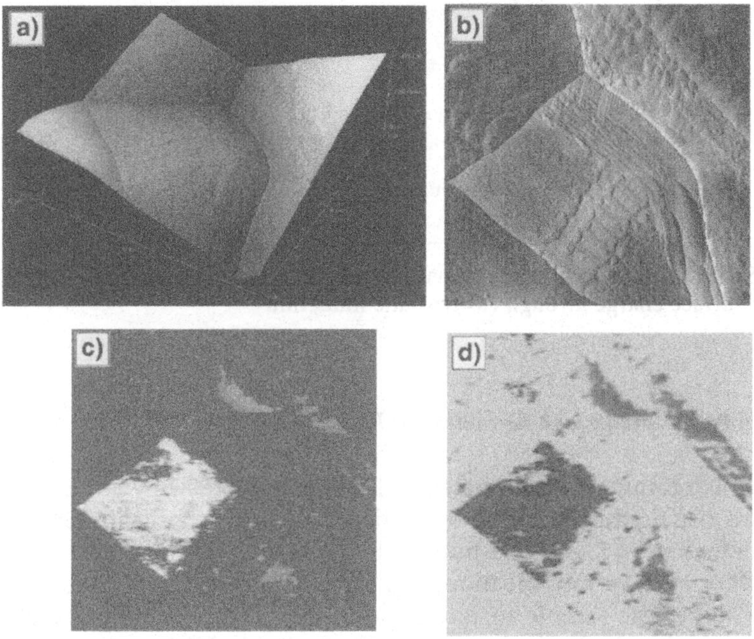

Fig. 43 **a** Topography (pseudo-3D-topography, height contrast 2.4 μm). **b** Normal force (error signal, a.u.). **c** A.c. current amplitude (contrast 4.18 μA). **d** Phase shift (contrast 90°). The excitation, superimposed to a d.c. signal of −10.57 V, is 4.4 V at 1 kHz. The scan width is 12 μm

slices are not conductive (with any bias) or whether they are simply not connected to the back electrode.

7.5 Materials Contrasts in the SFM Non-Contact Mode

A non-contact SFM with a metallised and grounded tip was used (Fig. 44); the lever resonance frequency f_{res} is ~75 kHz. The electrostatic induction in the tip, which depends on surface charge and on tip-sample distance, is measured with a synchronised lock-in amplifier. The alternating current reflects the amount of charge or the potential [455]. Also in this case the topography is measured simultaneously. This method provides results similar to Kelvin probe microscopy, where an external a.c. field vibrates the cantilever. Up to now the technique of current measurement leads to non-calibrated results, but it is much more simple and straightforward than Kelvin probe microscopy. The contrast depends on the dielectric constant, on the conductivity and on the work function (Fermi level). Several heterogeneous organic systems are presented.

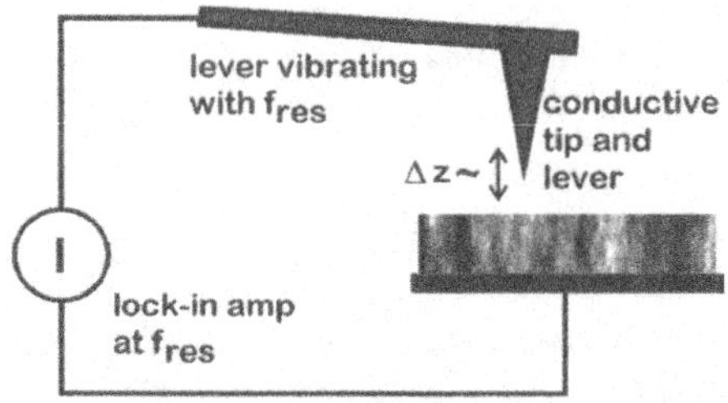

Fig. 44 Schematic diagram of the signal pathway in the non-contact mode SFM for imaging of surface charge through electrostatic induction

7.5.1
Industrial Polypropylene with Two Different Technical Additives

Additive containing polypropylene film was washed with ether. In the solvent, two usual additives were identified[7]: *cis*-13-docoseneamide[8] and tetrakis [methylene (3,5-di-*tert*-butyl-4-hydroxy-hydrocinnamate)] methane[9]. For further surface modification, e.g. for plasma treatment, it is essential to know whether the sample surface is pure polypropylene. In principle, cleaning procedures can be very effective, but most additives are specially designed to migrate back to the surface. Such a surface after Soxhlet cleaning with ether for several hours is shown in Fig. 45. After edge enhancing image processing, thin lines in the topography image (left) visualise small corrugation steps, corresponding to the white or the grey regions of two different surface charge levels (right image), which correlate to the two different organic contaminations. This occurs on all samples, depending on the washing procedure and reflects different concentration and distribution of the additive.

Surprisingly, the resolution is in the 10 nm range, so that about 10^{-17} g of this surface coverage can be detected. The assumption that after surface cleaning the additives migrate from the volume to the surface, e.g. to protect the foil from oxidation or electrostatic charge, can be confirmed.

7.5.2
Surface Charges on a Polymer Electret

A grounded metal grid with square holes (~8 μm) on a fluoroethylenepropylene film (FEP) was used for writing (duration 30 s) a pattern of surface charge by the beam of an electron microscope (typically 5–15 kV). The resulting accumulated charge is partially removed by washing with 2-propanol. The topography of FEP is shown in Fig. 46 (left). The bright squares are

[7] The two additives have been identified by matrix assisted laser desorption ionisation mass spectrometry (MALDI-MS) and attenuated total reflection infrared spectroscopy (ATR).

[8] Slip additive, Loxamid E, Henkel.

[9] Antioxidant agent, Irganox 1010, Ciba Geigy.

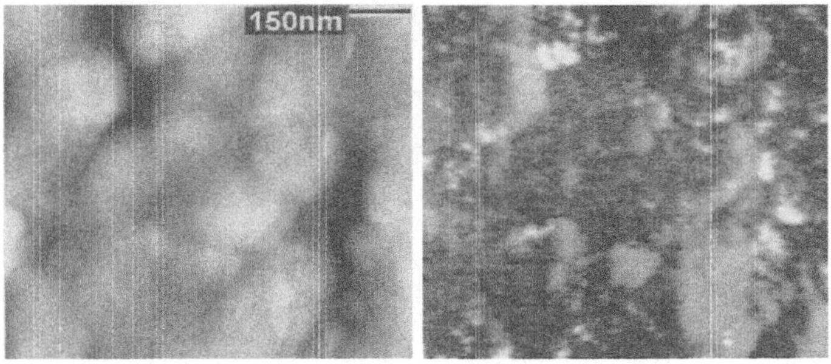

Fig. 45 Polypropylene film, contaminated with two different organic substances. *Left*: topography contrast with 8.1 nm corrugation (*shaded*). *Right*: surface charge contrast (a.u.)

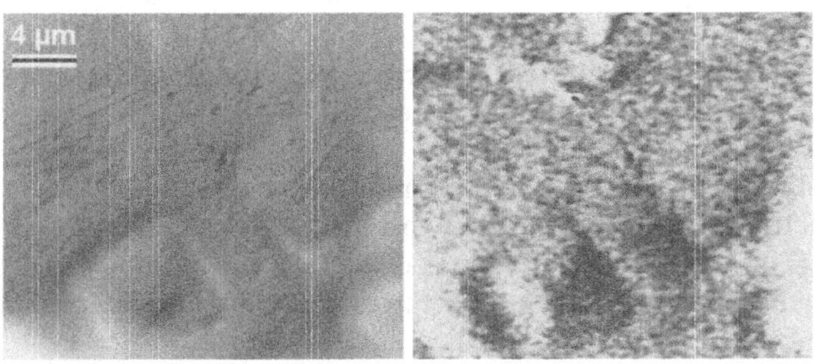

Fig. 46 Charged fluoroethylene-propylene (FEP). *Left*: topography contrast with 820 nm corrugation. *Right*: surface charge contrast (a.u.), grey scale equalised

due to the fact that the feedback keeps the sum of all force interactions (van der Waals and Coulomb forces) constant. The attractive force is so high that a pseudo-topographical superposition image is obtained [456]. However, calculations show that the amount of surface charges present in Figs. 45 and 47 leads to a negligible topography error of <1 nm. Due to the homogeneity of the surface, Fig. 46b is nearly featureless and reflects only the long-range charge interaction of the highly charged FEP surface.

7.5.3
Dewetting of an Incompatible Polymer Blend on a Gold Surface: Polyacryl-co-Styrene and Polybutadiene

A solution of poly(acryl-*co*-styrene)/polybutadiene blend (ABS) in tetrahydrofuran gives a thin, dewetted film on a gold substrate sputtered on a silicon wafer. The typical height of the steps in Fig. 47a is 90–100 nm. It is reasonable to suppose that a several nanometers thick film lies within the holes formed by the dewetting process. A free gold surface would differ dramatically from an insulator film, as can be demonstrated by an image of a scratch. The imaginary part of the charge induced displacement current,

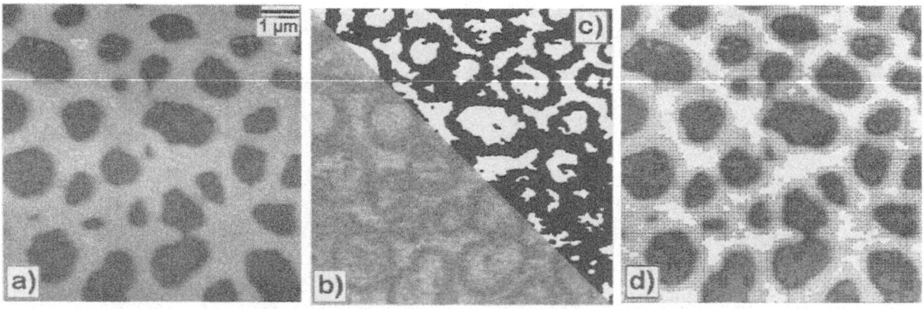

Fig. 47a–d Film of a poly(acryl-*co*-styrene)/polybutadiene blend (ABS), dissolved in tetrahydrofuran and dewetted on a gold substrate: **a** topography contrast, corrugation 172 nm; **b** surface charge contrast (a.u.); **c** surface charge contrast, two bit colour representation; **d** superposition of the topography and the two-bit image (*crosshatched*) of the surface charge, showing the heterogeneity of the bridges and the grooves

given in Fig. 47b, is attributed to carrier and dipole relaxation processes. The surface charge distribution is represented by Fig. 47c as a two-bit image. Dark areas belong to polybutadiene, whereas the bright sections are poly (acryl-*co*-styrene). This composition analysis is confirmed by measurements of local frictional properties. If the two-bit image is superimposed to the topography (Fig. 47a), the preferential dewetting morphology can be found: polybutadiene prefers to form the edges close to and within the deeper spots.

7.5.4
Liquid Crystalline Polysiloxane: Surfaces and Edges of a Smectic Layered Ferroelectric

Low molar mass ferroelectric liquid crystals have their application potential if short switching times in the electric field can be achieved. Even typical polymer or elastomer properties can be synthesised; however the coupling between the mesogenes and the polymer backbone must be minimised. The samples[10] used are liquid crystalline substances with a polysiloxane backbone. The principles of synthesis [457], their temperature dependent behaviour in an SFM [458] and the response on mechanical deformation [459] of this class of substances is described in detail in the literature. We are interested in understanding whether our method for the determination of surface charges is applicable to the special arrangement of dipole orientation in the sample of smectic layers which, in a coarse estimation, are perpendicular to the electric field probed by the conducting, vibrating SFM tip. Figure 48 presents the results and, due to the high risk of artefacts by using high amplification of the electrical signal, the scanning direction was changed twice and is indicated by the black and white arrows. Without so many simultaneously acquired signals it would be rather difficult to give a correct interpretation of the data. The edges of the smectic layers, marked by the white and black

[10] Courtesy of H.M. Brodowsky, Leipzig, Germany.

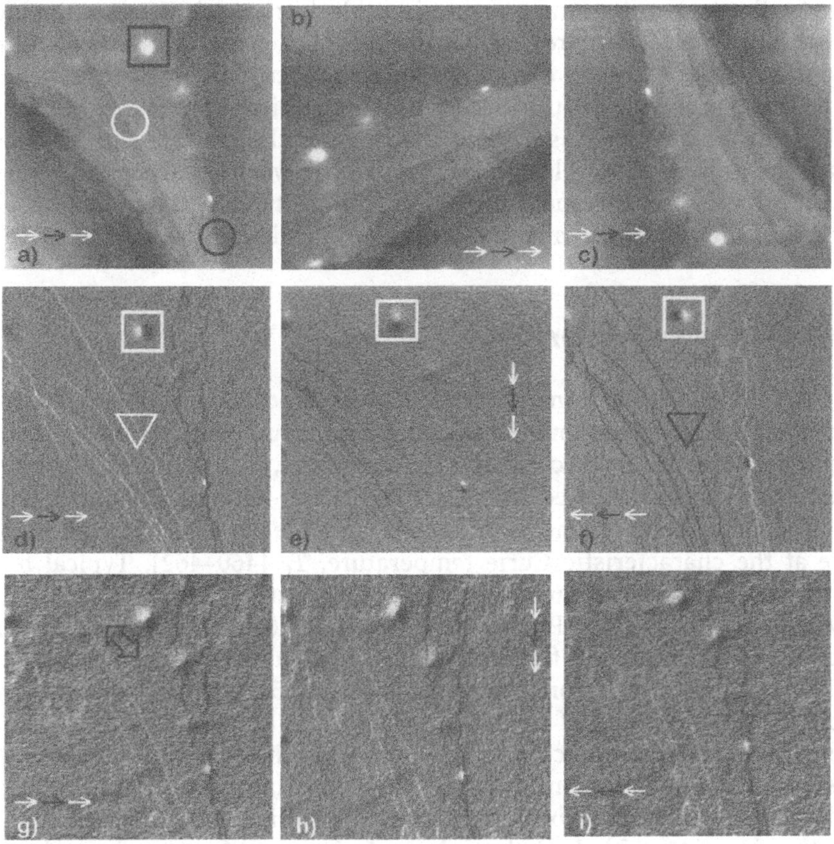

Fig. 48a–h Film of a ferroelectric liquid crystalline polysiloxane. Non-contact SFM images of: **a–c** topography; **d–f** error signal (force feedback); **g,h** a.c. current amplitude for three different scanning directions (0° in a, d and g, 90° in b, e and h, and 180° in c, f and i). The scan width in all images is 25 μm, the topography corrugation is 53 nm (a), 80 nm (b) and 72 nm (c); **d–f** are in a.u. The a.c. current contrast is about 10 pA at 67.61 kHz

circle, appear as differently charged (48 g). The left edges, where the tip has to climb 4–5 nm high steps, appear brighter in the surface charge image, i.e. the surface charge density is larger than that of the other regions. In the case of the right edges, the tip is descending and the corresponding image of the a.c. current shows a drop at the edges of the steps. The fact that the feedback is not able to keep the average tip-sample-distance constant at these steps (as clearly indicated by the lines in Fig. 48g) might be the cause of some artefacts. It is reasonable to assume that during the climbing the tip is closer to the surface, leading to a higher current amplitude (white triangle), and vice versa during the descending. However, the images of the scanning direction performed in 90° and 180° directions compared to the images of the first column demonstrate clearly, that this is not the case. Edges which have a higher surface charge are brighter in all the tree scanning directions even if the feedback error has an opposite sign (black triangle). Further, the blob-like structure (black square), where the feedback errors are larger and depend only on the scanning direction (white squares), appears very similar in all surface charge images (the direction of their orientation is marked by the double-headed arrow). It is evident that for the class of ferroelectric liquid crystalline polymers and elastomers it would be worth studying the properties on a local scale in much more detail, in particular the relaxation phenomena of their mechanical and electrical properties.

7.6 Piezoresponse Force Microscopy (PFM)

Ferroelectric materials consist of permanent dipoles in a non-centrosymmetric crystal structure and exhibit a spontaneous polarisation. If a sufficiently high electric field (coercitive field strength, E_c) is applied, the polarisation state can be switched between two thermo-dynamically stable states. The transition between the ferroelectric and the paraelectric phase takes place at the characteristic Curie temperature, T_C [460–462]. Typical ferroelectric materials are barium titanate ($BaTiO_3$), triglycine sulfate (TGS) or polyvinylidenedifluoride (PVDF) [460, 462]. Ferroelectrics are also piezo-/pyroelectric and therefore find various applications as sensors and actuators [463, 464]. Similar to ferromagnetic materials [465], ferroelectrics consist of a complex arrangement of domains. The domain structure and orientation determine the macroscopic switching properties and the piezo-/pyroelectric constants. Methods for imaging domains are important procedures for studying their size, distribution, switching kinetics and stability. Non-destructive imaging with high lateral resolution was first performed using SEM for WO_3 [466] and later on for PVDF [467, 468]. Birk et al. [469] were the first who measured *local* piezoelectric hysteresis loops by exciting a ferroelectric polymer with an external a.c. voltage and measured the periodic thickness variation with an STM. However, direct domain observation by means of STM is prevented because ferroelectric materials are insulators and need an additional conductive coating.

Domains with varying polarisation states differ in their physical properties. SPM based methods take advantage from differences in the optical properties (scanning near field optical microscopy (SNOM)) [470–472], the friction (friction force microscopy (FFM)) [473–475], the piezoelectric coefficients (PFM) [432, 434, 476–484], and the depolarising fields/surface potential (electrostatic force microscopy (EFM) or scanning surface potential microscopy (SSPM)) [456, 485–488]. In the following we will focus on PFM. Güthner and Dransfeld [489] were the first who used a home-built contact mode SFM with a conductive probe to polarise small domains with d.c. voltage pulses applied to the back electrode of the ferroelectric film. Subsequently, they imaged domains with an a.c. voltage on the tip during scanning. The oscillating electric field induces the converse piezoelectric effect and modulates the tip position normal to the surface with the excitation frequency. This modulation is extracted from the cantilever deflection signal by means of lock-in technique. The phase signal of the oscillation contains the information on domain polarity because the converse piezoelectric constant changes its sign when the spontaneous polarisation is switched [489]. The described experimental set-up is called piezoresponse force microscopy or, in alternative terms, "voltage modulated SFM" (VM-SFM) [483, 490, 491], or "dynamic contact mode electrical force microscopy" (DC-EFM) [492, 493].

In Sect. 7.3, Eqs. (18) and (19) describe the Maxwell stress forces acting on a conductive tip when a combined d.c./a.c. voltage is applied. For the PFM set-up we have to complete the total interaction force by the additional effects of piezoelectricity, electrostriction and the spontaneous polarisation. Both electromechanical effects cause an electric field-induced thickness variation and modulate the tip position. The spontaneous polarisation causes surface charges and changes the Maxwell stress force. If the voltage $U(t)=U_{DC}+U_{AC}*\sin(\omega t)$ is applied, the resulting total force $F_{total}(z)$ consists of three components (see also Eq. 19): F_{static}, F_ω and $F_{2\omega}$. F_{static} is the static cantilever deflection which is kept constant by the feedback loop. $F_{2\omega}$ contains additional information on electrostriction and Maxwell stress and will not be considered in detail here (for details see, e.g. [476]). The relevant component for PFM is F_ω [476, 477]:

$$F_\omega(z) = \left[-\frac{\partial C}{\partial z} \left(U_{DC} + \frac{P_{el}}{\varepsilon_0 \varepsilon_{ferro}} z_0 \right) + d_{33}\, k_{eff} + 2\frac{M_{33}}{z_0} U_{DC} k_{eff} \right] U_{AC} \sin(\omega t) \tag{20}$$

where C is the tip-sample capacitance, P_{el} the polarisation, z_0 the sample thickness, d_{33} the converse piezoelectric constant and M_{33} the electric field dependent electrostrictive constant. The index notation of the electromechanical constants denote the direction for the electric field and the resulting strain, respectively [494]. For the sake of simplicity the mechanical property of the tip/polymer contact is described by a spring with the effective force constant k_{eff} [6].

In Eq. (20) the three terms are related to the Maxwell stress (first), piezoelectric effect (second) and electrostriction (third). In order to obtain information about ferroelectricity via piezoresponse measurements, we need a link between the spontaneous polarisation and the piezoelectric constant. According to Furukawa and Damjanovic, piezoelectricity in ferroelectrics can be explained as electrostriction biased by the spontaneous polarisation if their paraelectric phase is nonpolar and centrosymmetric [461, 495, 496]. Therefore the d_{33} constant depends on the spontaneous polarisation P^s_{el}:

$$d_{33} = 2\,\varepsilon_{33}\, Q_{33}\, P^S_{el} \tag{21}$$

where Q_{33} is the polarisation dependent electrostrictive constant and ε_{33} the permittivity component. If the domain polarity is switched, Eq. (21) predicts a change in sign of d_{33}. Figure 49 illustrates the effect of a polarisation change from 0° to 180° on the tip oscillation phase. The measured PFM-signal is recorded either as amplitude B and phase Φ or as $B\cos\Phi$ (real part)

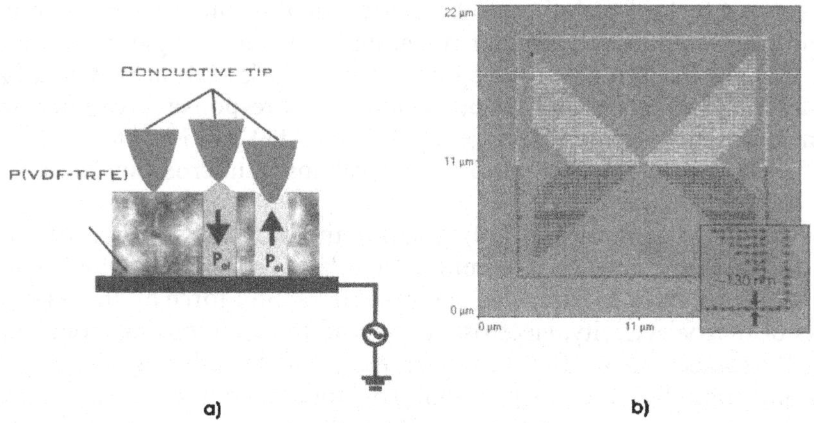

Fig. 49 Principle of domain contrast in PFM: **a** image shows the phase change of the tip oscillation for imaging a 0° and a 180° domain; **b** image is an example for the domain contrast ($B\sin\Phi$). The sample is a ~130 nm thick P(VDF-*co*-TrFE) film poled with voltage pulses at $U=\pm30$ V and a length of $\Delta\tau=500$ µs. The voltage polarity was changed from negative to positive after finishing the half of the scan. PFM imaging was done with 4 V at 92.33kHz

simultaneously with the topography. Both phase and real part contain the information about the domain polarity. In conclusion, it has to be emphasised that the observed thickness variation is always the result of an integration of d_{33} over the whole film thickness. Therefore PFM can only measure effective d_{33} values.

7.6.1 *Polarising Small Domains in Ferroelectric Polymer Films*

When the electric field is higher than the coercitive field strength, the spontaneous polarisation is switched and the dipoles reorientate along the field lines. This process of domain switching for $E_{el}=E_c$ can be described in three steps: i) nucleation of an anti-parallel domain, ii) domain growth and iii) saturation of the polarisation [461, 462].

For the application, e.g. as data storage media or as micro- and nanodevices, the domain sizes have to be reduced. The first poling experiments, performed by Schilling [468], reached domains sizes of ~500 nm in thin P(VDF-TrFE) films with a pulsed electron beam of a SEM. As already mentioned, Güthner and Dransfeld [489] demonstrated that poling of small domains by means of a PFM is possible down to lateral sizes of ~1 µm. For technological applications it is necessary to control precisely the domain position and to automate the process. A high repeatability of positioning the SFM tip is possible with a scanner based on a closed-loop x-y system as described in Sect. 6. Voltage pulses applied to the back electrode are trigged during a scan in the same way as for DPL. The pixel-based mask file includes the information for the domain location in the scan field. For each "high" bit a voltage pulse is triggered. The result of such a poling procedure for a ferroelectric copolymer (see Sect. 7.6.2) is shown in Fig. 49b for $\Delta\tau$=500 μs and U=±30 V. The voltage polarity was changed from positive to negative after half of the scan was finished. The scan speed per pixel should be sufficiently small against the pulse duration ($\Delta\tau$) to avoid domains with elliptical shape. This implies that high speed writing is only possible with ultra short pulses.

7.6.2 *Domain Growth Kinetics and Piezoelectric Hysteresis for Thin P(VDF-co-TrFE) Films*

Poly(vinylidene-*co*-trifluoroethylene) (75/25) is a statistical composition of the two monomer units (CH_2-CF_2) and (CHF-CF_2) in the ratio of 75/25, and the Curie temperature, T_C, is ~120–125 °C [497–500]. The sample crystallises both from solution and the melt in the ferroelectric all-trans configuration [462, 498]. The films are spin coated from a 1 wt% solution of methylethylketone on a gold coated glass slide and are subsequently annealed for 2 h at 135 °C to increase the crystallinity and the remanent polarisation [501]. The kinetics of domain growth is studied by means of rectangular voltage pulses of fixed height and different length. As an initial step, a large domain (50 μm×50 μm) is poled with a constant potential of U=+30 V during a slow scan (~0.5 Hz/line). After poling this background, a series of pulses with a constant potential of U=−30 V and lengths varying from 50 μs to 50 ms are applied. The domain area (A_d) growth follows a kinetics which is best fitted with a power function $A_d=A_o+\chi^*\tau^n$ (n<1), where A_o is the minimum domain area (Fig. 50a). It has to be pointed out that domain growth kinetics is strongly dependent on the local electric field distribution. Therefore a quantitative comparison between different samples is only possible when the same tip geometry is used.

Comparing the cross sections of piezoresponse images ($B\cos(\Phi)$) of domains grown in 50 μs, 5 ms and 50 ms (Fig. 50b), three sections of growth can be distinguished:

1. $\Delta\tau$=50 μs is the minimum pulse length necessary to form a domain, which is sufficiently stable to be imaged with PFM. Figure 50b indicates that the 50 μs domain does not reach the maximum piezoresponse at the centre and therefore it can not penetrate the entire film thickness of ~140 nm.
2. Between 50 μs and 5 ms, the domains expand in all three dimensions.
3. For $\Delta\tau$>5 ms, the maximum intensity is reached at the domain centre indicating that the whole film thickness is poled. Further growth only takes place in lateral dimensions.

As mentioned above, domain switching in ferroelectrics is accompanied by domain nucleation, moving domain walls and restructuring of dipoles and charges. A characteristic feature of this irreversible process is the appearance of a hysteresis loop in the dependence of dielectric displacement

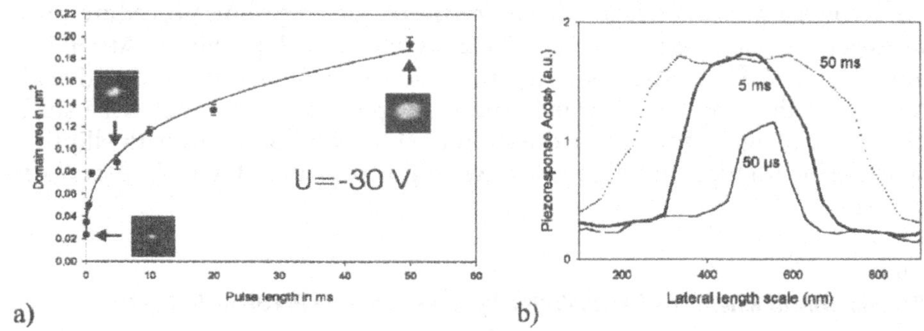

Fig. 50 Domain growth kinetics observed with PFM. Background poling prior to pulse poling was performed with U=+30 V. Rectangular voltage pulses of U=−30 V and length variations from 50 µs to 50 ms were applied to the back electrode. PFM imaging was done with 4 V at 92.33 kHz

D_{el} on the electric field strength. Figure 51 shows the schematic hysteresis loop for a macroscopic measurement. Increasing the electric field (starting from zero) causes a nonlinear increase in D_{el} until the saturated polarisation $P^{sat}{}_{el}$ is reached. Further increase of D_{el} is an effect of the permittivity [460]. Reversing the electric field, a finite remanent polarisation, $P^{r}{}_{el}$, is reached at zero field, which demonstrates the effect of spontaneous polarisation alignment during the first quarter of the cycle. If the electric field reaches the negative coercitive strength, $-E_c$, the polarisation switches in the antiparallel direction and saturates again for higher fields. Reversing the sweep direction once more, the negative remanent polarisation ($-P^{r}{}_{el}$) state is reached and afterwards the orientation switches at E_{el}=E_c. As described in Eq. (21), the correlation between the spontaneous polarisation and the piezoelectric con-

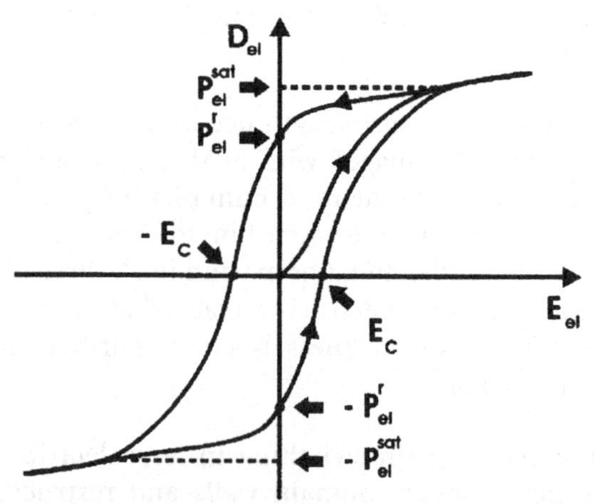

Fig. 51 Schematic macroscopic hysteresis loop of a ferroelectric material. Therein E_c denotes the coercitive field strength, $P^{r}{}_{el}$ the remanent polarisation and $P^{sat}{}_{el}$ the saturation polarisation

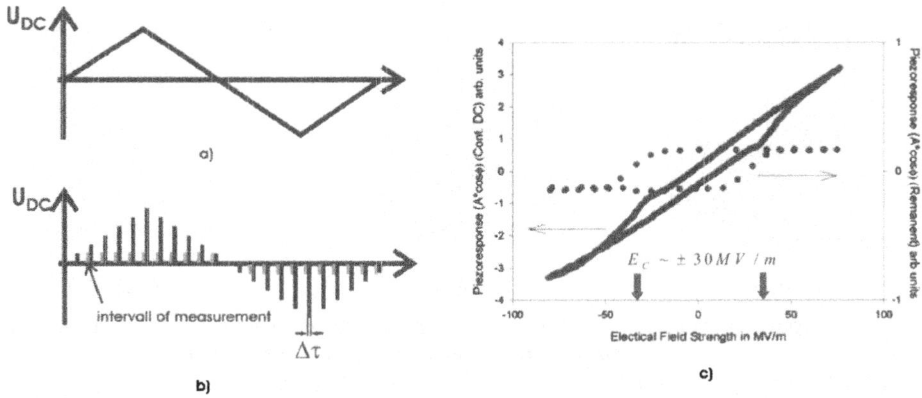

Fig. 52a–c Triangular voltage variation for: **a** *in-field*; **b** *remanent* loops. *Grey bars* in **b** represent the measurement interval where the PFM signal is recorded (100 ms in this experiment); c the resulting *in-field* loops (*straight line*) and *remanent* loops (*dotted line*) are shown. The pulse length for the remanent loop was 750 ms and piezoresponse was measured with 2 V at 90 kHz in both cases

stant d_{33} predicts that the spontaneous polarisation and the piezoresponse must show a similar hysteresis loop [493, 502]. Experiments on a local scale can be performed in two ways. In the first, the additional d.c. field is cycled linearly at a very low rate (<0.01 Hz) and exceeds $E_{c.}$ The PFM signal (ω-component) is continuously probed during the cycle and the loops are named *in-field* [503] or *continuous d.c.* loops [493] (Fig. 52a,c). In the second, a sequence of voltage pulses ($\Delta\tau$=750 ms) is applied and the piezoresponse is measured only at the end of a delay interval ($\Delta\tau$=750 ms) after the pulse is finished (Fig. 52b). The delay is necessary to give the system time for relaxation and establishing a steady state. Here the piezoresponse senses only the remanent polarisation P^r_{el} and is therefore named remanent [503] or pulse *d.c.* loop [493]. Equation (20) predicts a linear dependence of the PFM-signal on U_{DC} due to Maxwell stress and electrostriction. This strong influence on the loop shape is shown in Fig. 52c (continuous line). In contrast, the remanent loop lacks this linear superposition, because during PFM measurement the external field is switched off. Despite these differences, both curves provide similar values for the coercitive field strength E_c (~30 MV/m).

7.6.3
Poling DPL Fabricated Sub-μm Ferroelectric Polymer Structures

Spin coated copolymer films show a decrease in remanent polarisation if the film thickness [504, 505] decreases. The application of piezoelectric materials in micro-electro-mechanical systems (MEMS) or sensors makes it often necessary to decrease the lateral dimensions of the elements. Recently, Alexe et al. [506, 507] fabricated freestanding microcells with lateral dimensions down to ~100 nm and heights of ~110 nm from a ferroelectric PZT by direct

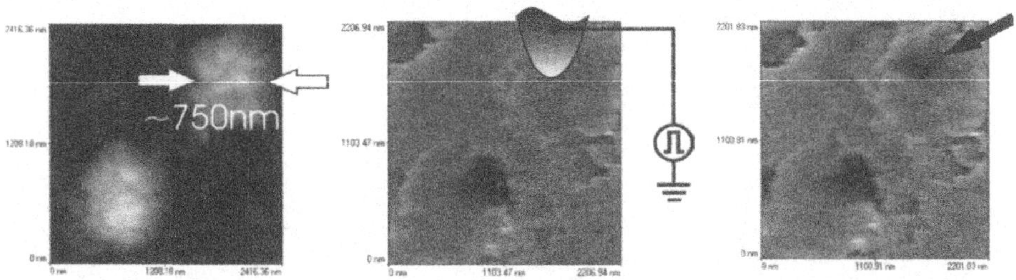

a) b) c)

Fig. 53 **a** Topography of two P(VDF-*co*-TrFE) nanostructures (height ~75 nm) fabricated with DPL. **b** Piezoresponse image of the same structures before a voltage pulse. **c** PFM image after the application of a voltage pulse (ΔU=−19 V; $\Delta\tau$=20 ms), PFM measured with 700 mV at 35 kHz. The newly formed domain is marked by an *arrow*. The lower left domain (*black dot* in **b**) was poled in a previous scan under the same conditions

electron beam writing. They used PFM for probing the local ferroelectric behaviour and found, e.g. shifts in the local hysteresis loops. The shifts were explained as a dimension effect caused by fixed polarisation states at the side-walls and at the electrode interface [508]. In Fig. 53, we show the first poling experiment of DPL-fabricated sub-μm structures of P(VDF-*co*-TrFE). The upper polymer structure (Fig. 53b) is poled with a voltage pulse ΔU=−19 V and $\Delta\tau$=20 ms. In the lower structure a domain was poled under the same conditions as in a previous scan. However, for these polymeric structures (lateral dimensions ~750 nm, thickness ~70 nm) no size effect has been found, yet.

PFM is limited by two factors: the local dimension of the probe and the sensitivity to the vertical tip movement. The smallest domains polarised in ferroelectric copolymer and subsequently imaged with PFM had diameters of ~25 nm [509]. A vertical sensitivity of ~1 pm is reasonable and implies that for thin films with low excitation voltages (~1 V) the smallest detectable d_{33} value is ~1 pm/V.

8 Concluding Remarks

Several applications of scanning force microscopy (SFM) and related techniques in polymer science have been given in the above sections. The reviewed results were gathered from surfaces of cross-sectioned bulk polymers, polymer-matrix composites, and polymer blends as well as free surfaces of polymer samples such as films, or surfaces prepared by means of replica techniques. The materials contrasts reported on range from several mechanical ones via thermal to electrical ones.

Owing to the very nature of SFM-type microscopes, scanning a sharp probe across a sample surface with the tip of the cantilever being extremely close to the sample surface, its most straightforward application is to rather flat and smooth samples exhibiting structures on small length scales, typically ranging from some nanometres to some tens of microns. Consequently, SFM has been widely applied to thin films and has boosted the research on phase separation phenomena occurring in thin films of polymer blends or copolymers [13]. Albeit the growing relevance of micro- and nanotechnologies as well as technologies for surface modification, there are plenty of promising applications of SFM to bulk polymer materials. The substitution of other materials by light polymers and polymer-matrix composites and the resulting issues to be studied demand valuable characterisation tools. Issues to be addressed may be related to composition or morphology of the polymer material, as well as to effects originating from processing or ageing. Effects arising at buried interfaces seem to be of paramount interest, since these interfaces may be the locus of initial mechanical stresses, physicochemical ageing effects, or charge-transfer processes. In particular, interfaces between macromolecules and metallic electrodes play a major role when making active devices up of organic molecules, e.g. in the case of organic light emitting diodes (oLED) or organic transistors. Thus, beyond the purely mechanical properties of the interphasial region, electrical metal-polymer interactions are of major concern such as electron injection (LEDs, transistors) or extraction (solar cells) [510].

Application of the surface technique SFM to buried interfaces requires sectioning of the samples. Depending on the technique for preparing the cross-sections and the mechanical properties of the polymer system to be investigated, the roughness of the cross-sectional surface may be rather high which in turn renders the SFM measurements on these surfaces difficult. Undoubtedly, free surfaces can be also very rough, for instance after tribological loading or after certain kinds of plasma treatment. In general, strong corrugations and pronounced roughness characteristics make it more difficult to keep constant the (mean) tip-sample distance. In particular, when measuring in mechanical contact of tip and sample, the contact area is ill-defined which may impair materials contrasts such as the electrical or thermal tip-sample current. Pronounced roughness characteristics and large scan areas require larger displacements than achievable with the usual piezoelectric elements and driving voltages. Moreover, the feedback-loop for adjusting the vertical cantilever position needs to act rapidly and as precise as possible, which in turn requires careful optimisation of the feedback-loop as well as fast electronics.

Other microscopy techniques such as light microscopy or scanning electron microscopy provide image ranges changeable over a similar number of length scales to those of SPM. Furthermore, for these microscopy techniques the time necessary for acquiring images is in general considerably shorter as compared to the one typical for SPM-type microscopes. For practical reasons (saving time, especially in industrial applications) as well as scientific ones (imaging of fast moving atoms or molecules), various approaches were proposed towards faster operation of SFM-type microscopes. However, when measuring tip-sample interactions exhibiting a low signal-to-noise ratio, some minimum sampling time may be necessary which in turn limits the acceptable scanning velocity. In general, high-speed scanning requires cantilevers with high resonance frequencies in order to ensure short time constants. Since cantilevers with a high resonance frequency need to be comparatively small and stiff, these requirements may collide with the needs resulting from their application to soft materials.

Rather than focusing solely on the development of fast-scanning SFM-type microscopes, the versatility and effectivity (as related to the acquisition time) may also be enhanced by hybrid-type approaches, i.e. by combining SFM-type microscopes with more rapid ones. Beyond the mere addition of the advantages of each of the single microscopes, even new (synergy) benefits may arise. For instance, when having combined an SFM-type microscope with a scanning electron microscope (SEM) [511, 512] within a single sample chamber, the sample surface can be imaged fast by means of the SEM, concentration profiles can be measured by means of the electron beam microprobe (EDX) technique, electron beam lithography (EBL) experiments can be performed, and last but not least, the various SFM-based contrasts can be measured on the same sample surface, without removing it from the sample chamber. For being able to operate at various environmental conditions rather than being limited to ultra-high vacuum (UHV) conditions, an environmental scanning electron microscope (ESEM) seems to be highly desirable. Furthermore, the combination of SEM with SFM renders possible the employment of lithography-techniques operating at different length scales. Whereas the small-scale wiring and the electrodes can be accomplished by means of EBL, at the nanometre scale the lithography steps can be performed by techniques based on SPM.

SFM-based materials contrasts can be thought of as a kind of micro-spectroscopy [513] providing both spatial resolution and materials sensitivity. In the more particular sense of micro-spectroscopy, measurement techniques such as Raman or infrared spectroscopy are being employed for mapping chemical properties, e.g. for recording molecular distribution maps. In general, at each pixel of the map (part of) a spectrum is being recorded and from this set of spectra a number of images can be displayed by selecting a certain value of the parameter under variation and giving the systems response at that parameter value. Similarly, in the special case of driving through the tip-sample distance, chemical forces are being measured. Accordingly, measuring force-distance curves is often referred to as (chemical) force spectroscopy [6]. In the more strict sense of spectroscopy, a frequency is being swept, e.g. the frequency of the light illuminated in case of scanning near-field optical microscopy (SNOM). Also the near-field fluorescence response may be investigated when dealing with fluorescent materials, possibly after dye-labeling. Essentially, the optical tip-sample interaction is achieved either by employing a fibre-tip aperture or an apertureless metal tip [513]. When being able to engage the same tip for other SFM-based modes, additional valuable information may be gained by combined measurement of SNOM-maps and SFM-based mechanical, thermal or electrical materials contrasts.

References

1. Binnig G, Quate CF, Gerber C (1986) Phys Rev Lett 56:930
2. Binnig G, Rohrer H (1982) Helv Phys Acta 55:726
3. Pohl DW, Denk W, Lanz M (1984) Appl Phys Lett 44:651
4. Lesko JJ, Jayaraman K, Reifsnider KL (1996) Key Engin Mat 116/117:61
5. Vettiger P, Despont M, Drechsler U, Dürig U, Häberle W, Lutwyche MI, Rothuizen HE, Stutz R, Widmer R, Binnig GK (2000) IBM J Res Develop 44:323
6. Cappella B, Dietler G (1999) Surface Sci Rep 34:1
7. Krüger D, Anczykowski B, Fuchs H (1997) Ann Phys-Leipzig 6(5):341
8. Magonov SN, Reneker DH (1997) Annu Rev Mater Sci 27:175
9. Maeda Y, Matsumoto T, Kawai T (1999) Appl Surf Sci 140:400
10. Ohnesorge FM (1999) Surf Interface Anal 27:379
11. Livshits AI, Shluger AL (1999) Appl Surf Sci 141:274
12. Moreno-Herrero F, dePablo PJ, Colchero J, Gómez-Herrero J, Baró AM (2000) Surf Sci 453:152
13. Sheiko SS (2000) Adv Pol Sci 151:61
14. Dubourg F, Couturier G, Aimé JP, Marsaudon S, Leclere P, Lazzaroni R, Salardenne J, Boisgard R (2001) Macromol Symp 167:177
15. Reiniger M, Basnar B, Friedbacher G, Schleberger M (2002) Surf Interface Anal 33:85
16. Anczykowski B, Krüger D, Fuchs H (1996) Phys Rev B 53:15,485
17. Spatz JP, Sheiko S, Möller M, Winkler RG, Reineker P, Marti O (1997) Langmuir 13:4699
18. Kühle A, Sørensen AH, Bohr J (1997) J Appl Phys 81:6562
19. Sarid D, Hunt JP, Workman RK, Yao X, Peterson CA (1998) Appl Phys A 66:S283
20. Haugstad G, Jones RR (1999) Ultramicroscopy 76:77
21. Behrend OP, Odoni L, Loubet JL, Burnham NA (1999) Appl Phys Lett 75:2551
22. Chen X, Davies MC, Roberts CJ, Tendler SJB, Williams PM, Burnham NA (2000) Surf Sci 460:292
23. Knoll A, Magerle R, Krausch G (2001) Macromolecules 34:4159
24. Paulo AS, García R (2001) Surf Sci 471:71
25. Akari SO, van der Vegte EW, Grim PCM, Belder GF, Koutsos V, ten Brinke G, Hadziioannou G (1994) Appl Phys Lett 65:1915
26. Nysten B, Legras R, Costa JL (1995) J Appl Phys 78:5953
27. Motomatsu M, Nie HY, Mizutani W, Tokumoto H (1996) Thin Solid Films 273:304
28. Kajiyama T, Tanaka K, Ge SR, Takahara A (1996) Progr Surf Sci 52:1
29. Labardi M, Allegrini M, Marchetti E, Sgarzi P (1996) J Vac Sci Technol B 14:1509
30. Vancso GJ, Allston TD, Chun I, Johansson LS, Liu G, Smith PF (1996) Int J Polym Anal Characterization 3:89
31. Jandt KD (1998) Mat Sci Eng R21:221
32. Cappella B, Baschieri P, Frediani C, Miccoli P, Ascoli C (1997) Nanotechnology 8:82
33. Rosa-Zeiser A, Weilandt E, Hild S, Marti O (1997) Meas Sci Technol 8:1333
34. Krotil HU, Stifter T, Waschipky H, Weishaupt K, Hild S, Marti O (1999) Surf Interface Anal 27:336
35. de Pablo PJ, Colchero J, Gómez-Herrero J, Baró AM (1998) Appl Phys Lett 73:3300
36. Maivald P, Butt HJ, Gould SAC, Prater CB, Drake B, Gurley JA, Elings VB, Hansma PK (1991) Nanotechnology 2:103
37. Colchero J, Luna M, Baró AM (1996) Appl Phys Lett 68:2896
38. Rabe U, Amelio S, Kester E, Scherer V, Hirsekorn S, Arnold W (2000) Ultrasonics 38:430
39. Florin EL, Radmacher M, Fleck B, Gaub HE (1994) Rev Sci Instrum 65:639
40. Han W, Lindsay SM, Jing T (1996) Appl Phys Lett 69:4111
41. Syed Asif SA, Wahl KJ, Colton RJ (1999) Rev Sci Instrum 70:2408
42. Troyon M, Wang Z, Pastre D, Lei HN, Hazotte A (1997) Nanotechnology 8:163
43. O'Shea SJ, Welland ME, Pethica JB (1994) Chem Phys Lett 223:336
44. Krotil HU, Weilandt E, Stifter T, Marti O, Hild S (1999) Surf Interface Anal 27:341

45. Burns AR, Carpick RW (2001) Appl Phys Lett 78:317
46. Overney RM, Meyer E, Frommer J, Güntherodt HJ, Fujihira M, Takano H, Gotoh Y (1994) Langmuir 10:1281
47. Bar G, Rubin S, Parikh AN, Swanson BI, Zawodzinski TA Jr, Whangbo MH (1997) Langmuir 13:373
48. Krotil HU, Stifter T, Marti O (2000) Rev Sci Instrum 71:2765
49. Magonov SN, Elings V, Papkov VS (1997) Polymer 38:297
50. Oulevey F, Burnham NA, Gremaud G, Kulik AJ, Pollock HM, Hammiche A, Reading M, Song M, Hourston DJ (2000) Polymer 41:3087
51. Grandy DB, Hourston DJ, Price DM, Reading M, Silva GG, Song M, Sykes PA (2000) Macromolecules 33:9348
52. Hammiche A, Pollock HM, Song M, Hourston DJ (1996) Meas Sci Technol 7:142
53. Fabian JH, Scandella L, Fuhrmann H, Berger R, Mezzacasa T, Musil C, Gobrecht J, Meyer E (2000) Ultramicroscopy 82:69
54. Fiege GBM, Altes A, Heiderhoff R, Balk LJ (1999) J Phys D Appl Phys 32:L13
55. Majumdar A (1999) Annu Rev Mater Sci 29:505
56. Pollock HM, Hammiche A (2001) J Phys D Appl Phys 34:R23–R53
57. Mazeran PE, Loubet JL (1997) Tribol Lett 3:125
58. Burnham NA, Kulik AJ, Gremaud G, Gallo PJ, Oulevey F (1996) J Vac Sci Technol B 14:794
59. Rabe U, Amelio S, Kopycinska M, Hirsekorn S, Kempf M, Göken M, Arnold W (2002) Surf Interface Anal 33:65
60. Dinelli F, Assender HE, Takeda N, Briggs GAD, Kolosov OV (1999) Surf Interface Anal 27:562
61. Drobek T, Stark RW, Gräber M, Heckl WM (1999) New J Phys 1:15.1
62. Cuberes MT, Assender HE, Briggs GAD, Kolosov OV (2000) J Phys D Appl Phys 33:2347
63. Rabe U, Janser K, Arnold W (1996) Rev Sci Instrum 67:3281
64. Yamanaka K, Nakano S (1998) Appl Phys A 66:S313
65. Turner JA, Hirsekorn S, Rabe U, Arnold W (1997) J Appl Phys 82:966
66. Rabe U, Kester E, Arnold W (1999) Surf Interface Anal 27:386
67. Dinelli F, Biswas SK, Briggs GAD, Kolosov OV (2000) Phys Rev B 61:13,995
68. Rabe U, Turner J, Arnold W (1998) Appl Phys A 66:S277
69. Claesson PM, Ederth T, Bergeron V, Rutland MW (1996) Adv Colloid Interface Sci 67:119
70. Cain RG, Page NW, Biggs S (2000) Phys Rev E 62:8369
71. Giesbers M, Kleijn JM, Fleer GJ, Stuart MAC (1998) Colloids Surf A 142:343
72. Ecke S, Butt HJ (2001) J Colloid Interface Sci 244:432
73. Laarz E, Meurk A, Yanez JA, Bergstrom L (2001) J Am Ceram Soc 84:1675
74. Drelich J, Nalaskowski J, Gosiewska A, Beach E, Miller JD (2000) J Adhes Sci Technol 14:1829
75. Biggs S, Spinks G (1998) J Adhes Sci Technol 12:461
76. Chaudhury MK, Weaver T, Hui CY, Kramer EJ (1996) J Appl Phys 80:30
77. Zauscher S, Klingenberg DJ (2000) J Colloid Interface Sci 229:497
78. Zauscher S, Klingenberg DJ (2001) Colloids Surf A 178:213
79. Dai H, Hafner JH, Rinzler AG, Colbert DT, Smalley RE (1996) Nature 387:147
80. Moloni K, Buss MR, Andres RP (1999) Ultramicroscopy 80:237
81. Cho JM, Sigmund WM (2002) J Colloid Interface Sci 245:405
82. Lee LH (1990) The chemistry and physics of solid adhesion. In: Lee LH (ed) Fundamentals of adhesion. Plenum Press, New York, p 1
83. Fröberg JC, Rojas OJ, Claesson PM (1999) Int J Miner Process 56:1
84. Taylor R, Williams RS, Chi VL, Bishop G, Fletcher J, Robinett W, Washburn S (1994) Surf Sci Lett 306:L534
85. Agrait N, Rodrigo JG, Rubio G, Sirvent C, Vieira S (1994) Thin Solid Films 253:199
86. Landman U, Luedtke WD, Ringer E (1992) Wear 153:3
87. Tsukruk VV (2001) Adv Mater 13:95

88. Komvopoulos K (1996) Wear 200:305
89. Maboudian R, Howe RT (1997) J Vac Sci Technol B 15:1
90. Kahn H, Heuer AH, Jacobs SJ (1999) Materials Today 2:3
91. Persson BNJ (1998) Sliding friction—physical principles and applications. Springer, Berlin Heidelberg New York, p 47
92. Bhushan B (2001) Wear 251:1105
93. Landman U, Luedtke WD, Burnham NA, Colton RJ (1990) Science 248:454
94. Maugis D (1997) Rev Metall 94:655
95. Dedkov GV (2000) Phys Stat Sol 179:3
96. Shull KR (2002) Mat Sci Engin R 36:1
97. Hertz H (1881) J Reine Angew Math 92:156 (Reproduced in: Miscellaneous papers by Heinrich Hertz (1896) Macmillan, London, p 146)
98. Johnson KL, Kendall K, Roberts AD (1971) Proc R Soc London A 324:301
99. Derjaguin BV, Muller VM, Toporov YP (1975) J Colloid Interface Sci 53:314
100. Pashley MD (1984) Colloids Surf 12:69
101. Unertl WN (1999) J Vac Sci Technol A 17:1779
102. Sneddon IN (1965) Int J Eng Sci 3:47
103. Muller VM, Yushenko VS, Derjaguin BV (1980) J Colloid Interface Sci 77:91
104. Maugis D (1992) J Colloid Interface Sci 150:243
105. Carpick RW, Ogletree DF, Salmeron M (1999) J Colloid Interface Sci 211:395
106. Piétrement O, Troyon M (2000) Tribology Lett 9:77
107. Piétrement O, Troyon M (2000) J Colloid Interface Sci 226:166
108. Johnson KL, Greenwood JA (1997) J Colloid Interface Sci 192:326
109. Greenwood JA (1997) Proc R Soc London A 453:1277
110. Christenson HK (1988) J Colloid Interface Sci 121:170
111. Maugis D, Gauthier-Manuel B (1994) J Adhesion Sci Technol 8:1311
112. Feng JQ (2001) J Colloid Interface Sci 238:318
113. Barthel E (1998) J Colloid Interface Sci 200:7
114. Lantz MA, O'Shea SJ, Welland ME, Johnson KL (1997) Phys Rev B 55:10776
115. Carpick RW, Agraït N, Ogletree DF, Salmeron M (1996) Langmuir 12:3334
116. Ting TCT (1966) J Appl Mech 33:845
117. Roberts AD, Thomas AG (1975) Wear 33:45
118. Schapery RA (1975) Int J Fracture 11:369
119. Maugis D, Barquins M (1978) J Phys D 11:1989
120. Greenwood JA, Johnson KL (1981) Philosoph Mag A 43:697
121. Hui CY, Baney JM, Kramer EJ (1998) Langmuir 14:6570
122. Giri M, Bousfield DB, Unertl WN (2000) Tribology Lett 9:33
123. Giri M, Bousfield DB, Unertl WN (2001) Langmuir 17:2973
124. Johnson KL (2000) In: Tsukruk VV, Wahl KJ (eds) Microstructure and microtribology of polymer surfaces. American Chemical Society, ACS Symposium Series, vol 741, p 24
125. Vakarelski IU, Toritani A, Nakayama M, Higashitani K (2001) Langmuir 17:4739
126. Deruelle M, Hervet H, Jandeau G, Leger L (1998) J Adhesion Sci Technol 12:225
127. Heuberger M, Dietler G, Schlapbach L (1996) J Vac Sci Technol B 14:1250
128. O'Shea SJ, Welland ME (1998) Langmuir 14:4186
129. Nyland LR, Maughan DW (2000) Biophys J 78:1490
130. Jayachandran R, Boyce MC, Argon A (1993) J Adhesion Sci Technol 7:813
131. Oliver WC, Pharr GM (1992) J Mater Res 7:1564
132. Fischer-Cripps AC (2000) Vacuum 58:569
133. Pharr GM, Oliver WC, Brotzen FR (1992) J Mater Res 7:13
134. Strojny A, Xinyun X, Tsou A, Gerberich WW (1998) J Adhesion Sci Technol 12:1299
135. Petzold M, Landgraf J, Füting M, Olaf JM (1995) Thin Solid Films 264:153
136. Kulkarni AV, Bhushan B (1996) Mater Lett 29:221
137. Briscoe BJ, Sebastian KS, Sinha SK (1996) Phil Mag A 74:1159
138. Briscoe BJ, Fiori L, Pelillo E (1998) J Phys D Appl Phys 31:2395
139. Chechenin NG, Bottiger J, Krog JP (1997) Thin Solid Films 304:70

140. Göken M, Kempf M, Bordenet M, Vehoff H (1999) Surf Interface Anal 27:302
141. Johnson KL (1997) Proc R Soc London A 453:163
142. Savkoor AR, Briggs GAD (1977) Proc R Soc London A 356:103
143. Lantz MA, O'Shea SJ, Hoole ACF, Welland ME (1997) Appl Phys Lett 70:970
144. Johnson KL (1985) Contact mechanics. Cambridge University Press, Cambridge
145. Carpick RW, Ogletree DF, Salmeron M (1997) Appl Phys Lett 70:1548
146. Scherer V, Arnold W, Bhushan B (1999) Surf Interface Anal 27:578
147. Wahl KJ, Stepnowski SV, Unertl WN (1998) Tribology Lett 5:103
148. Israelachvili JN, Chen YL, Yoshizawa H (1995) Relationship between adhesion and friction forces. In: Rimai DS, DeMejo LP, Mittal KL (eds) Fundamentals of adhesion and interfaces. VSP, Zeist
149. Homola AM, Israelachvili JN, McGuiggan PM, Gee ML (1990) Wear 136:65
150. Bowden FP, Tabor D (1964) The friction and lubrication of solids—part 2. Clarendon Press, Oxford
151. Carpick RW, Salmeron M (1997) Chem Rev 97:1163
152. Greenwood JA (1992) Contact of rough surfaces. In: Singer IL, Pollock HM (eds) Fundamentals of friction: macroscopic and microscopic processes. Kluver Academic Publishers, London, p 37
153. Schwarz UD, Zwörner O, Köster P, Wiesendanger R (1997) Phys Rev B 56:6987
154. Briscoe BJ, Evans DCB (1982) Proc R Soc London A 380:389
155. Putman C, Kaneko R (1996) Thin Solid Films 273:317
156. Meyer E, Lüthi R, Howald L, Bammerlin M, Guggisberg M, Güntherodt HJ (1996) J Vac Sci Technol B 14:1285
157. Singer IL (1994) J Vac Sci Technol A 12:2605
158. Colchero J, Meyer E, Marti O (1998) Friction on an atomic scale. In: Bushan B (ed) Handbook of nano/microtribology. CRC Press, Boca Raton
159. Porto M, Zaloj V, Urbakh M, Klafter J (2000) Tribology Lett 9:45
160. Granick S (1999) Phys Today 52:26
161. Tovstopyat-Nelip I, Hentschel HGE (2000) Phys Rev E 61:3318
162. Family F, Hentschel HGE, Braiman Y (2000) J Phys Chem B 104:3984
163. Behme G, Hesjedal T (2000) Appl Phys A 70:361
164. Braiman Y, Hentschel HGE, Family F, Mak C, Krim J (1999) Phys Rev E 59:R4737
165. Berman AD, Ducker WA, Israelachvili JN (1996) Langmuir 12:4559
166. Carlson JM, Batista AA (1996) Phys Rev E 53:4153
167. Yoshizawa H, Israelachvili JN (1993) J Phys Chem 97:11,300
168. Persson BNJ (1997) Phys Rev B 55:8004
169. Bouhacina T, Desbat B, Aimé JP (2000) Tribology Lett 9:111
170. Binggeli M, Mate CM (1995) J Vac Sci Technol B 13:1312
171. Schenk M, Füting M, Reichelt R (1998) J Appl Phys 84:4880
172. Piner RD, Mirkin CA (1997) Langmuir 13:6864
173. Xu L, Bluhm H, Salmeron M (1998) Surf Sci 407:251
174. Gil A, Colchero J, Gómez-Herrero J, Baró AM (2001) Ultramicroscopy 86:1
175. Herminghaus S, Fery A, Reim D (1997) Ultramicroscopy 69:211
176. Luna M, Colchero J, Baró AM (1999) J Phys Chem B 103:9576
177. Wei ZQ, Wang C, Bai CL (2000) Surf Sci 467:185
178. Wei Z, Wang C, Wang Z, Liu D, Bai C (2001) Surf Interface Anal 32:275
179. Miranda PB, Xu L, Shen YR, Salmeron M (1998) Phys Rev Lett 81:5876
180. Wilken R (1998) Ph.D. thesis, University of Potsdam
181. Holländer A, Klemberg-Sapieha JE, Wertheimer MR (1995) J Polym Sci A 33:2013
182. Wilken R, Holländer A, Behnisch J (1998) Macromolecules 31:7613
183. Painter R, Arakawa ET, Williams MW, Ashley JC (1980) Radiat Res 83:1
184. Wilken R, Holländer A, Behnisch J (1999) Surf Coat Tech 116/119:991
185. Munz M (2002) Zur nanomechanischen Charakterisierung der Interphase verstärkter Polymere, Forschungsbericht 241 der Bundesanstalt für Materialforschung (BAM), Wirtschaftsverlag NW Verlag für neue Wissenschaft GmbH, Bremerhaven

186. Munz M, Sturm H, Schulz E, Hinrichsen G (1998) Composites A 29:1251
187. Munz M, Sturm H, Schulz E (2001) German patent DE 197 28 357 C2
188. Lopez-Manchado M, Torre L, Kenny JM (2000) Rubber Chem Technol 73:694
189. Tomasetti E, Nysten B, Rouxhet PG, Poleunis C, Bertrand P, Legras R (1999) Surf Interface Anal 27:735
190. Munz M, Schulz E, Ernst E (2002) Unpublished results
191. Cakmak M, Cronin SW (2000) Rubber Chem Technol 73:753
192. Tomasetti E, Legras R, Henri-Mazeaud B, Nysten B (2000) Polymer 41:6597
193. Gedde UW (1995) Polymer physics. Chapman & Hall, London, p 239
194. Crank J (1975) The mathematics of diffusion. Clarendon Press, Oxford, p 12
195. Theocaris PS, Sideridis EP, Papanicolaou GC (1983) J Reinf Plast Comp 4:396
196. Stamm M, Schubert DW (1995) Ann Rev Mater Sci 25:325
197. Drzal LT (1986) Adv Polym Sci 75:1
198. Williams JG, Donnellan ME, James MR, Morris WL (1990) Mater Sci Eng A 126:305
199. Sottos NR, McCullough RL, Scott WR (1992) Compos Sci Technol 44:319
200. Garton A, Daly JH (1985) Polym Compos 6:195
201. Ko YS, Forsman WC, Dziemianowicz TS (1982) Polym Eng Sci 22:805
202. Racich JL, Koutsky JA (1977) Boundary layer in thermosets. In: Labana S (ed) Chemistry and properties of crosslinked polymers. Academic Press, New York, p 103
203. Fitzer E, Geigl KH, Hüttner W, Weiss R (1980) Carbon 18:389
204. Palmese GR, McCullough RL (1994) J Adhesion 44:29
205. Munz M, Sturm H, Schulz E (2000) Surf Interface Anal 30:410
206. Bhushan B (1999) Principles and applications of tribology. Wiley, New York, p 204
207. Sanford WM, McCullough RL (1990) J Polym Sci B 28:973
208. Kuchling H (1985) Taschenbuch der Physik. Harri Deutsch, Thun Frankfurt/Main
209. Hong SG, Wang TC (1994) J Appl Polym Sci 52:1339
210. O'Brien J, Cashell E, Wardell GE, McBrierty VJ (1976) Macromolecules 9:653
211. Bitsanis IA, Pan CJ (1993) Chem Phys 99:5520
212. Tsagaropoulos G, Eisenberg A (1995) Macromolecules 28:6067
213. Kosmas MK (1990) Macromolecules 23:2061
214. Pangelinan AB, McCullough RL, Kelley MJ (1994) J Polym Sci B 32:2383
215. Munz M, Schulz E, Sturm H (2002) Surf Interface Anal 33:100
216. Jang H, Kim SJ (2000) Wear 239:229
217. Ostermeyer G-P (1999) Phys Mesomechanics 6:25
218. Brandrup J, Immergut EH (1989) Polymer handbook, 3rd edn. Wiley, Chichester
219. Wolfrum J, Ehrenstein GW (1999) J Appl Polym Sci 74:3173
220. Meyer E, Overney RM, Dransfeld K, Gyalog T (1998) Nanoscience—friction and rheology on the nanometer scale. World Scientific, Singapore
221. Sturm H (1999) Macromol Symp 147:249
222. Sturm H, Schulz E, Munz M (1999) Macromol Symp 147:259
223. Singer IL (1992) Solid lubrication processes. In: Singer IL, Pollock HM (eds) Fundamentals of friction: macroscopic and microscopic processes. Kluwer Academic Publishers, London, p 237
224. De Carvalho G, Frollini E, Dos Santos WN (1996) J Appl Polym Sci 62:2281
225. Abd El-Rahman AA, Tahoon KK, El-Salam A, Abousehly M, Elwahab A, El-Sharkawy A (1996) J Thermal Anal 47:1719
226. Yang Y (1996) Thermal conductivity. In: Mark JE (ed) Physical properties of polymers handbook. Springer, Berlin Heidelberg New York, p 111
227. Marotta EE, Fletcher LS (1996) J Thermophys Heat Transfer 10:334
228. Parihar SK, Wright NT (1997) Int Commun Heat Mass Transfer 24:1083
229. Luo K, Shi Z, Varesi J, Majumdar A (1997) J Vac Sci Technol B 15:349
230. Colchero J, Storch A, Luna M, Gomez Herrero J, Baró AM (1998) Langmuir 14:2230
231. Piner RD, Mirkin CA (1997) Langmuir 13:6864
232. Barber JR (1967) Wear 10:155
233. Lee K, Barber JR (1993) Wear 160:237

234. Severin D, Dörsch S (2001) Wear 249:771
235. Miyake S (1995) Appl Phys Lett 67:2925
236. Magno R, Bennett BR (1997) Appl Phys Lett 70:1855
237. Fang TH, Weng CI, Chang JG (2000) Nanotechnology 11:181
238. Sumomogi T, Endo T, Kuwahara K, Kaneko R, Miyamoto T (1994) J Vac Sci Technol B 12:1876
239. Göbel H, von Blanckenhagen P (1995) J Vac Sci Technol B 13:1247
240. Tegen S, Kracke B, Damaschke B (1997) Rev Sci Instrum 68:1458
241. Fujihira M, Takano H (1994) J Vac Sci Technol B 12:1860
242. Garnaes J, Bjørnholm T, Zasadzinski JAN (1994) J Vac Sci Technol B 12:1839
243. Jin X, Unertl WN (1992) Appl Phys Lett 61:657
244. Sohn LL, Willett RL (1995) Appl Phys Lett 67:1552
245. Yamamoto S, Yamada H, Tokumoto H (1995) Jpn J Appl Phys 34:3396
246. Gorwadkar S, Vinogradov GK, Senda K, Morita S (1995) J Appl Phys 78:2242
247. Yano T, Nagahara LA, Hashimoto K, Fujishima A (1994) J Vac Sci Technol B 12:1596
248. Khurshudov A, Kato K (1995) J Vac Sci Technol B 13:1938
249. Nie HY, Motomatsu M, Mizutani W, Tokumoto H (1995) J Vac Sci Technol B 13:1163
250. Pingue P, Lazzarino M, Beltram B, Cecconi C, Baschieri P, Frediani C, Ascoli C (1997) J Vac Sci Technol B 15:1398
251. Avramescu A, Uesugi K, Suemune I (1997) Jpn J Appl Phys 36:4057
252. Bouchiat V, Esteve D (1996) Appl Phys Lett 69:3098
253. Jung TA, Moser A, Hug HJ, Brodbeck D, Hofer R, Hidber HR, Schwarz UD (1992) Ultramicroscopy 42/44:1446
254. Wendel M, Kühn S, Lorenz H, Kotthaus JP, Holland M (1994) Appl Phys Lett 65:1775
255. Wendel M, Irmer B, Cortes J, Kaiser R, Lorenz H, Kotthaus JP, Lorke A, Williams E (1996) Superlattices Microstruct 20:349
256. Cortes Rosa J, Wendel M, Lorenz H, Kotthaus JP, Thomas M, Kroemer H (1998) Appl Phys Lett 73:2684
257. Klehn B, Kunze U (1998) Superlattices Microstruct 23:441
258. Klehn B, Kunze U (1999) J Appl Phys 85:3897
259. Klehn B, Skaberna S, Kunze U (1999) Superlattices Microstruct 25:474
260. Skaberna S, Versen M, Klehn B, Kunze U, Reuter D, Wieck AD (2000) Ultramicroscopy 82:153
261. Cappella B, Sturm H (2001) Acta Microscopica 10:8
262. Villarrubia JS (1996) J Vac Sci Technol B 14:1518
263. Cappella B, Sturm H (2002) J Appl Phys 91:506
264. Heyde M, Rademann K, Cappella B, Geuss M, Sturm H, Spangenberg T, Niehus H (2001) Rev Sci Instrum 72:136
265. Bhushan B, Koinkar VN (1994) Appl Phys Lett 64:1653
266. Drechsler D, Karbach A, Fuchs H (1998) Appl Phys A 66:S825
267. Han Y, Schmitt S, Friedrich K (1999) Appl Composite Mater 6:1
268. Wiesauer K, Springholz G (2000) J Appl Phys 88:7289
269. Ward IM (1971) Mechanical properties of solid polymers. Wiley, London
270. Vieweg R, Esser F (1975) Polymethacrylate. Carl Hanser, Munich
271. Elias H-G (1981) Makromoleküle. Hüthig & Wepf Verlag, Basel
272. Cappella B, Sturm H, Schulz E (2002) J Adh Sci Technol 16(7):921
273. Burnham NA, Colton RJ (1989) J Vac Sci Technol A 7:2548
274. Lee LP, Berger SA, Liepmann D, Pruitt L (1998) Sens Actuators A (71):144
275. Heyderman LJ, Schift H, David C, Ketterer B, Auf der Maur M, Gobrecht J (2001) Microelectron Eng 57/58:375
276. Kim YS, Suh KY, Lee HH (2001) Appl Phys Lett 79(14):2285
277. Mizes HA, Loh KG, Miller RJD, Ahuja SK, Grabowski EF (1991) Appl Phys Lett 59:2901
278. Sasaki M, Hane K, Okuma S, Torii A (1995) J Vac Sci Technol B 13:350
279. Weisenhorn AL, Maivald P, Butt HJ, Hansma PK (1992) Phys Rev B 45:11,226
280. Fujihira M, Tani Y, Furugori M, Akiba U, Okabe Y (2001) Ultramicroscopy 86:63

281. Okabe Y, Akiba U, Fujihira M (2000) Appl Surf Sci 157:398
282. Freitas AM, Sharma MM (2001) J Colloid Interface Sci 233:73
283. Halbritter J (1999) Appl Phys A 68:153
284. Sarid D (1991) Scanning force microscopy with applications to electric, magnetic and atomic forces. Oxford University Press, New York
285. Denk W, Pohl DW (1991) Appl Phys Lett 59:2171
286. Hochwitz T, Henning AK, Levey C, Daghlian C, Slinkman J, Never J, Kaszuba P, Gluck R, Wells R, Zekarik J, Finch R (1996) J Vac Sci Technol B 14:440
287. Wittpahl V, Ney C, Behnke U, Mertin W, Kubalek E (1999) Microelectron Reliability 39:951
288. Balk LJ, Cramer RM (1999) Microelectron Eng 49:191
289. Bai C (1992) Scanning tunneling microscopy and its application. Springer, Berlin Heidelberg New York
290. Wiesendanger R, Güntherodt HJ (1992) (eds) Scanning tunneling microscopy II. Springer, Berlin Heidelberg New York
291. Hamers RJ (1996) J Phys Chem 100:13,103
292. Hasegawa Y, Jia JF, Inoue K, Sakai A, Sakurai T (1997) Surf Sci 386:328
293. Yu ET (1997) Chem Rev 97:1017
294. Song Z, Pascual JI, Conrad H, Horn K, Rust H-P (2001) Appl Phys A 72:S159
295. Zhou L, Ho PKH, Zhang PC, Li SFY, Xu GQ (1998) Appl Phys A 66:S643
296. Battiston FM, Bammerlin M, Loppacher C, Guggisberg M, Lüthi R, Meyer E, Eggimann F, Güntherodt HJ (1998) Appl Phys A 66:S49
297. Balk LJ, Maywald M (1994) Mater Sci Eng B 24:203
298. Uchihashi T, Fukano Y, Sugawara Y, Morita S, Nakano A, Ida T, Okada T (1994) Jpn J Appl Phys 33:L1562
299. O'Shea SJ, Atta RM, Murrell MP, Welland ME (1995) J Vac Sci Technol B 13:1945
300. Gaworzewski P, Roos B, Borngräber J, Höppner K, Höppner W, Henninger U (1996) J Vac Sci Technol B 14:373
301. Teuschler T, Mahr K, Miyazaki S, Hundhausen M, Ley L (1996) J Vac Sci Technol B 14:1268
302. Nxumalo JN, Shimizu DT, Thomson DJ (1996) J Vac Sci Technol B 14:386
303. Olbrich A, Ebersberger B, Boit C (1998) Appl Phys Lett 73:3114
304. Richter S, Cahen D, Cohen SR, Gartsman K, Lyakhovitskaya V, Manassen Y (1998) Appl Phys Lett 73:1868
305. Ando A, Miki K, Hasunuma R, Nishioka Y (2001) Appl Phys A 72:S223
306. Kocka J, Stuchlikova H, Stuchlik J, Rezek B, Svrcek V, Fojtik P, Pelant I, Fejfar A (2001) Solid Stata Phenom 80/81:213
307. Okada Y, Miyagi M, Akahane K, Iuchi Y, Kawabe M (2001) J Appl Phys 90:192
308. Porti M, Rodriguez R, Nafria M, Aymerich X, Olbrich A, Ebersberger B (2001) J Non-Cryst Solids 280:138
309. Porti M, Nafria M, Aymerich X, Olbrich A, Ebersberger B (2002) J Appl Phys 91:2071
310. Houze F, Meyer R, Schneegans O, Boyer L (1996) Appl Phys Lett 69:1975
311. Farley AN (1997) Appl Phys Lett 70:3618
312. Morita S, Ishizaka T, Sugawara Y, Okada T, Mishima S, Imai S, Mikoshiba N (1998) Jap J Appl Phys 28:L1634
313. Schneegans O, Houze F, Meyer R, Boyer L (1998) IEEE T Compon Pack A 21:76
314. Houze F, Meyer R, Schneegans O, Boyer L (1997) Appl Phys Lett 70:3619
315. Gallo PJ, Kulik AJ, Burnham NA, Oulevey F, Gremaud G (1997) Nanotechnology 8:10
316. Macpherson JV, de Mussy JPG, Delplancke JL (2001) Electrochem Solid State 4:E33
317. Müller F, Müller A-D, Kröll M, Schmid G (2001) Appl Surf Sci 171:125
318. Liau YH, Scherer NF, Rhodes K (2001) J Phys Chem B 105:3282
319. Gu JH, Tanaka S, Otsuka Y, Tabata H, Kawai T (2002) Appl Phys Lett 80:688
320. Lee H, Maeda Y, Tanaka S, Tabata H, Tanaka H, Kawai T (2001) J Korean Phys Soc 39: S345
321. Amrein M, Schenk M, von Nahmen A, Sieber M, Reichelt R (1995) J Microsc 179:261

322. Wiegrabe W, Knapp HF, Eberhart H, Gatz R, Hartmann T, Heim M, Lorek C, Guckenberger R (1995) Rev Sci Instrum 66:4124
323. Yano K, Kuroda R, Shimada Y, Shida S, Kyogaku M, Matsuda H, Takimoto K, Egushi K, Nakagiri T (1996) J Vac Sci Technol B 14:1353
324. Guessab S, Boyer L, Houze F, Noel S, Schneegans O (2001) Synthetic Met 118:121
325. Planes J, Houze F, Chretien P, Schneegans O (2001) Appl Phys Lett 79:2993
326. Yagi A, Tsukada S, Takahashi Y, Morita S, Sugawara Y, Okada T, Imai S, Mikoshiba N (1990) J Vac Sci Technol A 8:336
327. Sturm H, Schulz E (1995) Acta Polym 46:379
328. Luo EZ, Xu JB, Wu W, Wilson IH, Zhao B, Yan X (1998) Appl Phys A 66:S1171
329. Kazimierski P, Lehmberg H (1998) Surf Coat Technol 98:939
330. Zhang L, Sakai T, Sakuma N, Ono T (2001) Diamond Relat Mater 10:829
331. De Pablo PJ, Gomez-Navarro C, Martinez MT, de Benito AM, Maser WK, Colchero J, Gomez-Herrero J, Baró AM (2002) Appl Phys Lett 80:2421
332. Knite M, Teteris V, Polyakov B, Erst D (2002) Mater Sci Eng C 19:15
333. Thomson RE, Moreland J (1995) J Vac Sci Technol B 13:1123
334. Lányi S, Török J, Rehurek P (1996) J Vac Sci Technol B 14:892
335. Lantz MA, O'Shea SJ, Welland ME (1998) Rev Sci Instrum 69:1757
336. Oyama Y, Majima Y, Iwamoto M (1999) Rev Sci Instrum 86:7087
337. Efimov A, Cohen SR (2000) J Vac Sci Technol A 18:1051
338. Hantschel T, Slesazeck S, Niedermann P, Eyben P, Vandervorst W (2001) Microelectron Eng 57/58:749
339. De Wolf P, Stephenson R, Trenkler T, Clarysse T, Hantschel T, Vandervorst W (2000) J Vac Sci Technol B 18:361
340. Hantschel T, Niedermann P, Trenkler T, Vandervorst W (2000) Appl Phys Lett 76:1603
341. Chen KW, Yu YH, Lei YM, Cheng LL, Sundaraval B, Luo EZ, Wong SP, Wilson IH, Chen LZ, Ren CX, Zou SC (2001) Appl Surf Sci 184:178
342. Eyben P, Xu M, Duhayon N, Clarysse T, Callewaert S, Vandervorst W (2002) J Vac Sci Technol B 20:471
343. Clarysse T, Caymax M, Vandervorst W (2002) Appl Phys Lett 80:2407
344. Kleinknecht HP, Sandercock JR, Meier H (1988) Scanning Microsc 2:1839
345. Martin Y, Abraham DW, Wickramasinghe HK (1988) Appl Phys Lett 52:1103
346. Williams CC, Hough WP, Rishton SA (1989) Appl Phys Lett 55:203
347. Neubauer G, Erickson A, Williams CC, Kopanski JJ, Rodgers M, Adderton D (1996) J Vac Sci Technol B 14:426
348. Leong J-K, McMurray J, Williams CC, Stringfellow GB (1996) J Vac Sci Technol B 14:3113
349. Nakagiri N, Yamamoto T, Sugimura H, Suzuki Y (1996) J Vac Sci Technol B 14:887
350. Kopanski JJ, Marchiando JF, Rennex BG (2000) J Vac Sci Technol B 18:409
351. Palmer RC, Denlinger EJ, Kawamoto H (1982) RCA Rev. 43:104
352. Born A, Wiesendanger R (1998) Appl Phys A 66:S421
353. Tomiye H, Yao T (2000) Appl Surf Sci 159:210
354. Tran T, Oliver DR, Thomson DJ, Bridges GE (2001) Rev Sci Instrum 72:2618
355. Stern JE, Terris BD, Mamin HJ, Rugar D (1988) Appl Phys Lett 53:2717
356. Terris BD, Stern JE, Rugar D, Mamin HJ (1989) Phys Rev Lett 63:2669
357. Terris BD, Stern JE, Rugar D, Mamin HJ (1990) J Vac Sci Technol A 8:374
358. Schönenberger C, Alvarado SF (1991) Mod Phys Lett B 5:871
359. Schönenberger C, Alvarado SF (1990) Phys Rev Lett 65:3162
360. Schönenberger C (1992) Phys Rev B 45:3861
361. Uchihashi T, Nakano A, Ida T, Andoh Y, Kaneko R, Sugawara Y, Morita S (1997) Jpn J Appl Phys I 36:3755
362. Wintle HJ (1997) Meas Sci Technol 8:508
363. Lord Kelvin (1898) Philos Mag 46:82
364. Zisman WA (1932) Rev Sci Instrum 3:367

365. Steinke R, Hoffmann M, Böhmisch M, Eisenmenger J, Dransfeld K, Leiderer P (1997) Appl Phys A 64:19
366. Nonnenmacher M, O'Boyle M, Wickramasinghe HK (1992) Ultramicroscopy 42:268
367. Mäckel R, Baumgärtner H, Ren J (1993) Rev Sci Instrum 64:694
368. Tanimoto M, Vatel O (1996) J Vac Sci Technol B 14:1547
369. Kikukawa A, Hosaka S, Imura R (1996) Rev Sci Instrum 67:1463
370. Semenikhin OA, Jiang L, Iyoda T, Hashimoto K, Fujishima A (1997) Electrochim Acta 42:3321
371. Boneberg J, Burmeister F, Schäfle C, Leiderer P, Reim D, Fery A, Herminghaus S (1997) Langmuir 13:7080
372. Lü J, Guggisberg M, Lüthi R, Kubon M, Scandella L, Gerber C, Meyer E, Güntherodt H-J (1998) Appl Phys A 66:S273
373. Jacobs HO, Leuchtmann P, Homan OJ, Stemmer A (1998) J Appl Phys 84:1168
374. O'Boyle MP, Hwang TT, Wickramasinghe HK (1999) Appl Phys Lett 74:2641
375. Baikie ID, Petermann U, Speakman A, Lagel B, Dirscherl KM, Estrup PJ (2000) J Appl Phys 88:4371
376. Takahashi T, Kawamukai T (2000) Ultramicroscopy 82:63
377. Ludeke R, Cartier E (2001) Appl Phys Lett 78:3998
378. Sturm H, Stark W, Schulz E (1995) German patent DE 195 32 838 A1
379. Inoue T, Yokoyama H (1994) Thin Solid Films 243:399
380. Itoh J, Nazuka Y, Kanemaru S, Inoue T, Yokoyama H (1996) J Vac Sci Technol B 14:2105
381. Eng LM, Abplanalp M, Günter P (1998) Appl Phys A 66: S679
382. Bugg CD, King PJ (1988) J Phys E 21:147
383. Williams CC, Slinkman J, Hough WP, Wickramasinghe HK (1989) Appl Phys Lett 55:1662
384. Abraham DW, Williams CC, Slinkman J, Wickramasinghe HK (1991) J Vac Sci Technol B 9:703
385. Barrett RC, Quate CF (1992) Ultramicroscopy 42/44:262
386. Leng Y, Williams CC (1994) Colloids Surf A 93:335
387. Yokoyama H, Inoue T (1994) Thin Solid Films 242:33
388. Mueller U, Hofchen S, Boehm C, Sprengepiel J, Kubalek E, Beyer A (1996) Microelectr Eng 31:235
389. Leyk A, Kubalek E (1996) Microelectr Eng 31:187
390. Goto K, Hane K (1997) Rev Sci Instrum 68:120
391. Nie HY, Horiuchi K, Yamauchi Y, Masai J (1997) Nanotechnology 8:A24
392. Belaidi S, Lebon F, Girard P, Leveque G, Pagano S (1998) Appl Phys A 66:S239
393. Franke K, Huelz H, Weihnacht M (1998) Surf Sci 416:59
394. Weaver JMR, Wickramasinghe KH (1991) J Vac Sci Technol B 9:1562
395. Müller F, Müller A-D, Hietschold M, Kämmer S (1998) Meas Sci Technol 9:734
396. Nie H-Y, Masai J (1998) Appl Phys A 66:S1059
397. Weaver JMR, Abraham DW (1991) J Vac Sci Technol B 9:1559
398. Völker M, Krieger W, Walther H (1993) J Appl Phys 74:5426
399. Yokohama H, Inoue T, Itoh J (1994) Appl Phys Lett 65:3143
400. Said RA, Bridges GE, Thomson DJ (1994) Appl Phys Lett 64:1442
401. Sprengepiel J, Boehm C, Kubalek E (1996) Microelect Eng 31:181
402. Böhm C, Sprengepiel J, Otterbeck M, Kubalek E (1996) J Vac Sci Technol B 14:842
403. Arnold L, Krieger W, Walther H (1997) Appl Phys Lett 51:786
404. Miyatani T, Horii M, Rosa A, Fujihira M, Marti O (1997) Appl Phys Lett 71:2632
405. Miyatani T, Okamoto S, Rosa A, Marti O, Fujihira M (1998) Appl Phys A 66:S349
406. Doris BB, Hegde RI (1995) Appl Phys Lett 67:3816
407. Ronning C, Wondratschek O, Büttner M, Hofsäss H, Zimmermann J, Leiderer P, Boneberg J (2001) Appl Phys Lett 79:3053
408. Luo K, Shi Z, Lai J, Majumdar A (1996) Appl Phys Lett 68:325
409. Mills G, Zhou H, Midha A, Donaldson L, Weaver JMR (1998) Appl Phys Lett 72:2900

410. Wang D, Tsau L, Wang KL, Chow P (1995) Appl Phys Lett 67:1295
411. Snow ES, Juan WH, Pang SW, Campbell PM (1995) Appl Phys Lett 66:1729
412. Park SW, Soh HT, Quate CF, Park S-I (1995) Appl Phys Lett 67:2415
413. Teuschler T, Mahr K, Miyazaki S, Hundhausen M, Ley L (1995) Appl Phys Lett 66:2499
414. Kado H, Tohda T (1995) Appl Phys Lett 66:2961
415. Koyanagi H, Hosaka S, Imura R, Shirai M (1995) Appl Phys Lett 67:2609
416. Quate CF (1997) Surf Sci 386:259
417. Avramescu A, Ueta A, Uesugi K, Suemune I (1998) Appl Phys Lett 72:716
418. Abadal G, Pérez-Murano F, Barniol N, Aymerich X (1998) Appl Phys A 66:S791
419. Wilder K, Quate CF, Adderton D, Bernstein R, Elings V (1998) Appl Phys Lett 73:2527
420. Kim BI, Pi UH, Khim ZG, Yoon S (1998) Appl Phys A 66:S95
421. Peterson CA, Ruskell TG, Pyle JL, Workman RK, Yao X, Hunt JP, Sarid D, Parks HG, Vermeire B (1998) Appl Phys A 66:S729
422. Avouris P, Martel R, Hertel T, Sandstrom R (1998) Appl Phys A 66:S659
423. Oesterschulze E (1998) Appl Phys A 66:S3
424. Legrand B, Stievenard D (1999) Appl Phys Lett 74:4049
425. Abadal G, Boisen A, Davis ZJ, Hansen O, Grey F (1999) Appl Phys Lett 74:2306
426. Lemeshko S, Gavrilov S, Shevyakov V, Roschin V, Solomatenko R (2001) Nanotechnology 12:273
427. Hu J, Xiao X, Salmeron M (1995) Appl Phys Lett 67:476
428. Hu J, Carpick RW, Salmeron M, Xiao X (1996) J Vac Sci Technol B 14:1341
429. Dai Q, Hu J, Freedman A, Robinson G, Salmeron M (1996) J Phys Chem 100:9
430. Hosaka S, Koyanagi H, Kikukawa A, Miyamoto M, Imura R, Ushiyama J (1995) J Vac Sci Technol B 13:1307
431. Gruverman A, Kolosov O, Hatano J, Takahashi K, Tokumoto H (1995) J Vac Sci Technol B 13:1095
432. Gruverman A, Auciello O, Tokumoto H (1998) Annu Rev Mater Sci 28:101
433. Takata K, Miki H, Kushida-Abdelghafar K, Torii K, Fujisaki Y (1998) Appl Phys A 66:S441
434. Christman JA, Woolcott RR Jr, Kingon AI, Nemanich RJ (1998) Appl Phys Lett 73:3851
435. Eng LM, Güntherodt H-J, Schneider GA, Kopke U, Saldana JM (1999) Appl Phys Lett 74:233
436. Sturm H, Geuss M, Schulz E (1999) Scanning force microscopy investigations on electroactive polymer films. In: Kosta AA (ed) Tenth International Symposium on Electrets (ISE10). IEEE, Piscataway, p 465
437. Likodimos V, Orlik XK, Pardi L, Labardi M, Allegrini M (2000) J Appl Phys 87:443
438. Date M, Kutani M, Sakai S (2000) J Appl Phys 87:863
439. Leinhos T, Stopka M, Oesterschulze E (1989) Appl Phys A 66:S65
440. Trenkler T, Hantschel T, Stephenson R, De Wolf P, Vandervorst W, Hellemans L, Malave A, Buchel D, Oesterschulze E, Kulisch W, Niedermann P, Sulzbach T, Ohlsson O (2000) J Vac Sci Technol B 18:418
441. Lantz MA, O'Shea SJ, Welland ME (1997) Phys Rev B 56:15,345
442. Lányi S (1999) Surf Interface Anal 27:348
443. Sturm H, Schulz E (1996) Composites Part A 27:677
444. Sturm H, Stark W, Bovtoun V, Schulz E (1996) Methods for simultaneous measurements of topography and local electrical properties using scanning force microscopy. In: Shanghai, Xia Z, Zhang H (eds) Ninth International Symposium on Electrets (ISE9). IEEE, Piscataway, p 223
445. Bovtoun VP, Sturm H, Pashkov VM (1996) Electroceramics V. Aveiro, Portugal
446. Sturm H, Schulz E (1995) Beitr Elektronenm Direktabb Oberfl 28:51
447. Wiesendanger R, Anselmetti D (1992) STM on layered materials. In: Güntherodt HJ, Wiesendanger R (eds) Scanning tunneling microscopy I. Springer, Berlin Heidelberg New York, p 131
448. Waiblinger M, Sommerhalter C, Pietzak B, Krauser J, Mertesacker B, Lux-Steiner MC, Klaumünzer S, Weidinger A, Ronning C, Hofsäß H (1999) Appl Phys A 69:239

449. Gröning O, Küttel OM, Gröning P, Schlapbach L (1997) Appl Surf Sci 111:135
450. Amaratunga GAJ, Silva SRP (1996) Appl Phys Lett 68:2529
451. Zhu W, Kochanski GP, Jin S, Seibles L (1995) J Appl Phys 78:2707
452. Krauser J, Hoffmann V, Harneit W, Waiblinger M, Stolterfoht N, Weidinger A, Trautmann C, Hofsäss H, Ronning C, Schultrich B, Sturm H (2001) AIP Conf Proc 591:507
453. Schulz E, Kalinka G, Auersch W (1996) J Macromol Sci-Phys 335:527
454. Kruse A, Schlett V, Baalmann A, Hennecke M (1993) Fresenius J Anal Chem 346:284
455. Bovtoun V, Sturm H, Leshchenko MA, Yakimenko YI (1997) Ferroelectrics 190:161
456. Lüthi R, Haefke H, Meyer K-P, Meyer E, Howald L, Güntherodt H-J (1993) J Appl Phys 74:7461
457. Brehmer M, Zentel R (1995) Macromol Rapid Comm 16:659
458. Brodowsky HM, Boehnke U-C, Kremer F (1997) Langmuir 13:5378
459. Brodowsky HM, Boehnke U-C, Kremer F (1999) Langmuir 15:274
460. Lines ME, Glass AM (1977) Principles and applications of ferroelectrics and related materials. Clarendon Press, Oxford
461. Damjanovic D (1998) Rep Prog Phys 61:1267
462. Nalwa S (1995) Ferroelectric polymers. Wiley, New York
463. Wang TT, Herbert JM, Glass AM (1989) The application of ferroelectric polymers. Blackie, New York
464. Busch-Vishniac IJ (1999) Electromechanical sensors and actuators. Springer, Berlin Heidelberg New York
465. Hubert A, Schäfer R (1998) Magnetic domains. Springer, Berlin Heidelberg New York
466. Le Bihan R, Sella C (1970) J Phys Soc Jpn 28:377
467. Koshida N, Yoshida S (1983) Jpn J Appl Phys 22:1744
468. Schilling D (1988) PhD-thesis, University of Konstanz
469. Birk H, Glatz-Reichenbach J, Li-Jie, Schreck E, Dransfeld K (1991) J Vac Sci Technol B 9:1162
470. Eng LM, Güntherodt H-J (2000) Ferroelectrics 236:35
471. Moyer PJ, Walzer K, Hietschold M (1995) Appl Phys Lett 87:2129
472. McDaniel EB, Hsu JWP (1996) J Appl Phys 80:1085
473. Correia A, Massanell J, Garcia N (1996) Appl Phys Lett 68:2796
474. Yang TJ, Mohideen U, Gupta MC (1997) Appl Phys Lett 71:1960
475. Tolstikhina L, Belugina NV, Shikin SA (2000) Ultramicroscopy 82:149
476. Franke K, Weihnacht M (1995) Ferroelectric Lett 19:25
477. Franke K (1995) Ferroelectric Lett 19:35
478. Franke K, Hülz H (1996) Ferroelectric Lett 21:93
479. Franke K, Hülz H, Seifert S (1997) Ferroelectric Lett 23:1
480. Franke K, Hülz H, Weihnacht M (1998) Surf Sci 415:178
481. Durkan C, Welland ME, Chu DP, Migliorat P (1999) Phys Rev B 60:16,198
482. Eng LM, Bammerlin M, Loppacher C, Guggisberg M, Bennewitz R, Lüthi R, Meyer E, Huser T, Heinzelmann H, Güntherodt H-J (1999) Ferroelectrics 222:153
483. Labardi M, Likodimos V, Allegrini M (2000) Phys Rev B 61:14,390
484. Okino H, Teruhiko I, Ebihara H, Yamada H, Matshushige K, Yamamoto T (2001) Jpn J Appl Phys 40:5828
485. Saurenbach F, Terris BD (1990) Appl Phys Lett 56:1703
486. Lüthi R, Haefke H, Gutmannsbauer W, Meyer E, Howald L, Güntherodt HJ (1994) J Vac Sci Technol B 12:2451
487. Ohigami J, Sugawara Y, Morita S, Nakamura E, Ozaki T (1996) Jpn J Appl Phys 35:2734
488. Kalinin SV, Bonnell D (2001) Phys Rev B 63:125,411
489. Güthner P, Dransfeld K (1992) Appl Phys Lett. 61:1137
490. Labardi M, Likodimos V, Allegrini M (2001) Appl Phys A 72:S79
491. Likodimos V, Labardi M, Allegrini M (2000) Phys Rev B 61:14,440
492. Hong JW, Noh KH, Park SI, Kwun SI, Khim ZG (1998) Phys Rev B 58:5078
493. Hong S, Woo J, Shin H, Jeon JU, Pak YE, Colla EL, Setter N, Kim E, Kwangsoo N (2001) J Appl Phys 89:1377

494. Nye JF (1985) Physical properties of crystals. Oxford Sci, New York
495. Furukawa (1989) IEEE Trans Electr Insul 24:375
496. Furukawa T, Seo N (1990) Jpn J Appl Phys 29:675
497. Furukawa T, Date M, Fukada E, Tajitsu Y, Chiba A (1980) Jpn J Appl Phys 19:L109
498. Furukawa T(1989) Phase Trans 18:143
499. Yamada T, Ueda T, Kitayama T (1981) J Appl Phys 52:948
500. Higashihata Y, Sako J, Yagi T (1981) Ferroelectrics 32:85
501. Tajitsu Y, Ogura H, Chiba A, Furukawa T (1987) Jpn J Appl Phys 26:554
502. Eng LM (1999) Nanotechnology 10:405
503. Abplanalp M (2001) PhD thesis, ETH Zurich
504. Tajitsu Y (1995) Jpn J Appl Phys 35:5418
505. Zhang QM, Xu H, Fei F, Cheng ZY, Feng X, You H (2001) J Appl Phys 89:2613
506. Alexe M, Harnagea C, Hesse D, Gösele U (1999) Appl Phys Lett 75:1793
507. Alexe M, Harnagea C, Erfurth W, Hesse D, Gösele U (2000) Appl Phys A 70:247
508. Alexe M, Harnagea C, Hesse D, Gösele U (2001) Appl Phys Lett 79:242
509. Matsushige K, Yamada H, Tanaka H, Horiuchi T, Chen XQ (1998) Nanotechnology 9:208
510. Fahlman M, Salaneck WR (2002) Surf Sci 500:904
511. Troyon M, Lei HN, Wang ZH, Shang GY (1997) Microsc Microanal Microstruct 8:393
512. Joachimsthaler I, Heiderhoff R, Balk LJ (2003) Meas Sci Technol 14:87
513. Dragnea B, Leone SR (2001) Int Rev Phys Chem 20(1):59

Editor: H. H. Kausch

Received: September 2002

Author Index Volumes 101–164

Author Index Volumes 1-100 see Volume 100

Subject Index

The manufacturer's authorised representative in the EU is Springer
Nature Customer Service Centre GmbH, Europaplatz 3, 69115 Heidelberg,
Germany. If you have any concerns regarding our products, please
contact ProductSafety@springernature.com

Printed and bound by CPI Group (UK) Ltd, Croydon, CR0 4YY
07/01/2026
02031056-0007